ERRATUM

Rising Above the Gathering Storm, Revisited
Rapidly Approaching Category 5
The National Academies Press, Washington, D.C.
ISBN-13: 978-0-309-16097-1; ISBN-10: 0-309-16097-9

On page 29, Table 2-1, replace Congressional Action for Recommendation B-5 with the following:

$415 million appropriated in the FY2009 Omnibus and ARRA.

RISING ABOVE
THE GATHERING STORM, REVISITED

Rapidly Approaching Category 5

By Members of the 2005 "Rising Above the Gathering Storm" Committee

Prepared for the Presidents of the

National Academy of Sciences

National Academy of Engineering

Institute of Medicine

NATIONAL ACADEMY OF SCIENCES,
NATIONAL ACADEMY OF ENGINEERING, AND
INSTITUTE OF MEDICINE
OF THE NATIONAL ACADEMIES

THE NATIONAL ACADEMIES PRESS
Washington, D.C.
www.nap.edu

THE NATIONAL ACADEMIES PRESS • **500 Fifth Street, N.W.** • **Washington, DC 20001**

Support for this project was provided by the National Academy of Sciences, the National Academy of Engineering, and the Institute of Medicine. Any opinions, findings, conclusions, or recommendations expressed in this publication are those of the author(s) and do not necessarily reflect the views of the organizations or agencies that provided support for the project.

International Standard Book Number-13: 978-0-309-16097-1 (Book)
International Standard Book Number-10: 0-309-16097-9 (Book)
International Standard Book Number-13: 978-0-309-16098-8 (PDF)
International Standard Book Number-10: 0-309-16098-7 (PDF)

Additional copies of this report are available from the National Academies Press, 500 Fifth Street, N.W., Lockbox 285, Washington, DC 20055; (800) 624-6242 or (202) 334-3313 (in the Washington metropolitan area); Internet, http://www.nap.edu

THE NATIONAL ACADEMIES
Advisers to the Nation on Science, Engineering, and Medicine

The **National Academy of Sciences** is a private, nonprofit, self-perpetuating society of distinguished scholars engaged in scientific and engineering research, dedicated to the furtherance of science and technology and to their use for the general welfare. Upon the authority of the charter granted to it by the Congress in 1863, the Academy has a mandate that requires it to advise the federal government on scientific and technical matters. Dr. Ralph J. Cicerone is president of the National Academy of Sciences.

The **National Academy of Engineering** was established in 1964, under the charter of the National Academy of Sciences, as a parallel organization of outstanding engineers. It is autonomous in its administration and in the selection of its members, sharing with the National Academy of Sciences the responsibility for advising the federal government. The National Academy of Engineering also sponsors engineering programs aimed at meeting national needs, encourages education and research, and recognizes the superior achievements of engineers. Dr. Charles M. Vest is president of the National Academy of Engineering.

The **Institute of Medicine** was established in 1970 by the National Academy of Sciences to secure the services of eminent members of appropriate professions in the examination of policy matters pertaining to the health of the public. The Institute acts under the responsibility given to the National Academy of Sciences by its congressional charter to be an adviser to the federal government and, upon its own initiative, to identify issues of medical care, research, and education. Dr. Harvey V. Fineberg is president of the Institute of Medicine.

The **National Research Council** was organized by the National Academy of Sciences in 1916 to associate the broad community of science and technology with the Academy's purposes of furthering knowledge and advising the federal government. Functioning in accordance with general policies determined by the Academy, the Council has become the principal operating agency of both the National Academy of Sciences and the National Academy of Engineering in providing services to the government, the public, and the scientific and engineering communities. The Council is administered jointly by both Academies and the Institute of Medicine. Dr. Ralph J. Cicerone and Dr. Charles M. Vest are chair and vice chair, respectively, of the National Research Council.

www.national-academies.org

2005 "RISING ABOVE THE GATHERING STORM" COMMITTEE MEMBERS PARTICIPATING IN "THE GATHERING STORM, REVISITED"[1]

NORMAN R. AUGUSTINE [NAE/NAS] (Chair) is the retired chairman and CEO of the Lockheed Martin Corporation and a former Undersecretary of the Army. He is a recipient of the National Medal of Technology.

CRAIG BARRETT [NAE] is retired chairman and CEO of Intel Corporation.

GAIL CASSELL [IOM] is vice president for scientific affairs and a Distinguished Lilly Research Scholar for Infectious Diseases at Eli Lilly and Company. She is the former president of the American Society for Microbiology and former member of the Food and Drug Administration Science Board and Advisory Committees to the Director of the National Institutes of Health and the Center for Disease Control.

NANCY GRASMICK is the Maryland state superintendent of schools.

CHARLES HOLLIDAY JR. [NAE] is the retired chairman of the Board and CEO of DuPont.

SHIRLEY ANN JACKSON [NAE] is president of Rensselaer Polytechnic Institute. She is a past president of the American Association for the Advancement of Science and was chairman of the U.S. Nuclear Regulatory Commission.

ANITA K. JONES [NAE] is University Professor Emerita at the University of Virginia. She served as director of defense research and engineering at the U.S. Department of Defense and was vice-chair of the National Science Board.

RICHARD LEVIN is president of Yale University and the Frederick William Beinecke Professor of Economics.

C. D. (DAN) MOTE JR. [NAE] is president emeritus of the University of Maryland and the Glenn L. Martin Institute Professor of Engineering.

---

[1]Additional members of the 2005 Committee:

STEVEN CHU [NAS], a Nobel Laureate in physics, is currently serving as U.S. Secretary of Energy.

ROBERT GATES, former president of Texas A&M University, is currently serving as U.S. Secretary of Defense.

JOSHUA LEDERBERG [NAS], recipient of the Nobel Prize in physiology/medicine, passed away on February 2, 2008.

CHERRY MURRAY [NAS/NAE] is dean of the School of Engineering and Applied Science at Harvard University. She is immediate past president of the American Physical Society and a past deputy director for science and technology at Lawrence Livermore National Laboratory. She was formerly a senior vice president at Bell Labs, Lucent Technologies.

PETER O'DONNELL JR. is president of the O'Donnell Foundation of Dallas, a private foundation that develops and funds model programs designed to strengthen engineering and science education and research.

LEE R. RAYMOND [NAE] is the retired chairman of the Board and CEO of Exxon Mobil Corporation.

ROBERT C. RICHARDSON [NAS] is the F. R. Newman Professor of Physics and the vice provost for research at Cornell University. He was a co-winner of the Nobel Prize in physics in 1996.

P. ROY VAGELOS [NAS/IOM] is the retired chairman and CEO of Merck & Co., Inc.

CHARLES M. VEST [NAE] is president of the National Academy of Engineering and is president emeritus of MIT and a professor of mechanical engineering. He is a recipient of the National Medal of Technology.

GEORGE M. WHITESIDES [NAS/NAE] is the Woodford L. & Ann A. Flowers University Professor at Harvard University. He has served as an adviser for the National Science Foundation and the Defense Advanced Research Projects Agency.

RICHARD N. ZARE [NAS] is the Marguerite Blake Wilbur Professor of Natural Science at Stanford University. He was chair of the National Science Board from 1996 to 1998.

"Gentlemen, we have run out of money. It is time to start thinking."

Sir Ernest Rutherford, Nobel Laureate (Physics)

Foreword

We are pleased to present this report authored by members of the committee that produced the 2005 report, *Rising Above the Gathering Storm: Energizing and Employing America for a Brighter Economic Future.*[1] We requested this new report to get the perspective of the original committee on progress and change since the 2005 report.

BACKGROUND

Rising Above the Gathering Storm was prepared in response to a request by a bipartisan group of Senators and Members of Congress who asked the National Academies to respond to the following questions:

> *What are the top 10 actions, in priority order, that federal policymakers could take to enhance the science and technology enterprise so that the United States can successfully compete, prosper, and be secure in the global community of the 21st century? What strategy, with several concrete steps, could be used to implement each of those actions?*[2]

These questions were posed in the context of rapid and deep changes in the global economy, investment patterns, advancing science and technology, and the global redistribution of skilled workforces, education, and innovation-driven industries. Moreover, there was widespread unease about long-term trends in U.S. investments in research, develop-

---

[1] NAS/NAE/IOM, *Rising above the Gathering Storm: Energizing and Employing America for a Brighter Economic Future,* National Academies Press, 2007. The initial report release was in 2005, with the final, edited book issued in 2007.

[2] Letters from Senators Jeff Bingaman and Lamar Alexander, dated May 27, 2005, and Congressmen Sherwood Boehlert and Bart Gordon, to NAS President Bruce Alberts.

ment and higher education, and special and deepening concern about the competitiveness of U.S. businesses and the state of the primary and secondary education attained by vast numbers of our children. *Rising Above the Gathering Storm* was drafted by a group of 20 distinguished Americans including then current or former corporate CEOs; university presidents; scientists, including three Nobel Laureates; philanthropists, former government officials; and education leaders.[3] Norman R. Augustine, retired CEO of Lockheed Martin and former Under Secretary of the Army, chaired the committee. A vast relevant literature was reviewed, updated, and summarized; a diverse group of 66 stakeholders was convened to help frame and contextualize the issues; and the committee formed consensus on its recommendations. Peers drawn from many relevant backgrounds and experiences reviewed the report prior to its release.

The original report informed the debate in Congress and within two presidential administrations, and, together with other reviews of America's competitive position and innovation environment, led to the passage with strong bipartisan support of the America COMPETES Act of 2007[4] that has formed the basis for debating and structuring federal policy and budgets, and prompted a great deal of activity at local, state, and regional levels as well.

THE CURRENT REVIEW

In the five years that have passed since *Rising Above the Gathering Storm* was issued, much has changed in our nation and world. Despite the many positive responses to the initial report, including congressional hearings and legislative proposals, America's competitive position in the world now faces even greater challenges, exacerbated by the economic turmoil of the last few years and by the rapid and persistent worldwide advance of education, knowledge, innovation, investment, and industrial infrastructure. Indeed the governments of many other countries in Europe and Asia have themselves acknowledged and aggressively pursued many of the key recommendations of *Rising Above the Gathering Storm,* often more vigorously than has the U.S. We also sense that in the face of so many other daunting near-term challenges, U.S. government and industry are letting the crucial strategic issues of U.S. competitiveness slip below the surface.

---

[3] The Committee on Prospering in the Global Economy of the 21st Century: An Agenda for American Science and Technology was authorized under the auspices of the NAS/NAE/IOM Committee on Science, Engineering, and Public Policy (COSEPUP). Its overall charge was to address cross-cutting issues in science and technology policy that affect the health of the national research enterprise.

[4] America Creating Opportunities to Meaningfully Promote Excellence in Technology, Education, and Science Act, Public Law 110-69, August 9, 2007.

For these reasons, we believed that the nation would be well served by an update of the global context and events since the original report. We therefore asked Mr. Augustine, assisted by National Academies staff, to prepare a first draft of this update document and then work with the available members of the original committee to refine it. Each of the available members of the committee generously agreed to do so as a matter of national service. The resulting report was then anonymously peer reviewed by ten individuals with a wide range relevant expertise and experience. The results of this process are reported herein and have the unanimous support of the available members of the 2005 committee.[5]

As presidents of the National Academy of Sciences, National Academy of Engineering, and Institute of Medicine, we are pleased to convey this report to interested readers. We believe that it will serve to inform the public and policy makers, rekindle and advance an urgent national dialogue, and stimulate further strong and sustained bipartisan effort to ensure the future competitiveness, innovation capacity, economic vitality, and job creation in the opening decades of this century.

Ralph J. Cicerone
President, National
Academy of Sciences

Charles M. Vest
President, National
Academy of Engineering

Harvey V. Fineberg
President, Institute
of Medicine

---

[5] One member, Joshua Lederberg, is now deceased. Steven Chu is currently serving as U.S. Secretary of Energy and Robert Gates is currently serving as U.S. Secretary of Defense and therefore they could not participate.

PREFACE

During the summer of 2005, the National Academy of Sciences, the National Academy of Engineering and the Institute of Medicine undertook a study of America's evolving competitiveness in the global economy. The study resulted in a 500-page volume that became known as the *"Gathering Storm"* report. It focused upon the ability of Americans to compete for employment in a job market that increasingly knows no geographic boundaries.

The Executive Summary of the original report began, "The United States takes deserved pride in the vitality of its economy, which forms the foundation of our high quality of life, our national security, and our hope that our children and grandchildren will inherit ever-greater opportunities." But the report concluded that, "Without a renewed effort to bolster the foundations of our competitiveness, we can expect to lose our privileged position." Contained in the initial report were twenty specific actions that were intended to help assure that America could in fact remain competitive.

Five years have passed since the initial report was prepared, a period in which a great deal has changed...and a great deal has not changed. The recommendations included several actions that relate specifically to the physical sciences and engineering. Reflecting evolving federal budget priorities, the present report also briefly considers the biological sciences, which after a period of growth have begun to see their funding erode. This document, unanimously approved by participating members of the original *Gathering Storm* committee, revisits and updates the earlier findings.

CONTENTS

Executive Summary

In 2005, bipartisan requests from the United States House of Representatives and the United States Senate prompted the National Academies to conduct a study of America's competitiveness in the newly evolved global marketplace. An Academies committee comprised of twenty individuals of highly diverse professional backgrounds, supported by the staff of the Academies and many others, subsequently conducted a review of America's competitive position and released a report that has become popularly referred to as the "*Gathering Storm*" report after the first line in its title.

The Academies' review culminated in four overarching recommendations, underpinned by twenty specific implementing actions. Generally strong bipartisan support was granted these findings on Capitol Hill and in the White House and a number of the recommendations were eventually implemented. However, the preponderance of the enabling financial resources was provided in the American Recovery and Reinvestment Act ("Stimulus Legislation") which is presumed to be a one-time, albeit two-year, initiative. Similarly, the Authorizing Legislation to implement many of the *Gathering Storm* recommendations, known as the America COMPETES Act, was specified to expire after three years; i.e., in the 2010 fiscal year.

Although significant progress has been made as a result of the above legislation, the *Gathering Storm* effort once again finds itself at a tipping point. It is widely agreed that addressing America's competitiveness challenge is an undertaking that will require many years if not decades; however, the requisite federal funding of much of that effort is about to terminate. In order to sustain the progress that has begun it will be necessary to (1) reauthorize the America COMPETES Act, and (2) "institutionalize" funding and oversight of the *Gathering Storm* recommendations—

or others that accomplish the same purpose—such that funding and policy changes will routinely be considered in future years' legislative processes.

It would be impossible not to recognize the great difficulty of carrying out the *Gathering Storm* recommendations, such as doubling the research budget, in today's fiscal environment...with worthy demand after worthy demand confronting budgetary realities. However, it is emphasized that actions such as doubling the research budget are *investments* that will need to be made if the nation is to maintain the economic strength to provide for its citizens healthcare, social security, national security, and more. One seemingly relevant analogy is that a non-solution to making an over-weight aircraft flight-worthy is to remove an engine.

The original *Gathering Storm* competitiveness report focuses on the ability of America and Americans to compete for jobs in the evolving global economy. The possession of quality jobs is the foundation of a high quality life for the nation's citizenry.

The report paints a daunting outlook for America if it were to continue on the perilous path it has been following in recent decades with regard to sustained competitiveness.

The purpose of the present report is to assess changes in America's competitive posture in the five years that have elapsed since the *Gathering Storm* report was initially published and to assess the status of implementation of the National Academies' recommendations.

Robert Solow received a Nobel Prize in economics in part for his work that indicated that well over half of the growth in United States output per hour during the first half of the twentieth century could be attributed to advancements in knowledge, particularly technology.[1] This period was, of course, before the technology explosion that has been witnessed in recent decades. The National Academies *Gathering Storm* committee concluded that a primary driver of the future economy and concomitant creation of jobs will be *innovation*, largely derived from advances in science and engineering. While only four percent of the nation's work force is

---

[1] R.M. Solow, "Technical Change and the Aggregate Production Function." *Review of Economics and Statistics*, 39: 312-320, 1957.

composed of scientists and engineers, this group disproportionately creates jobs for the other 96 percent.[2]

When scientists discovered how to decipher the human genome it opened entire new opportunities in many fields including medicine. Similarly, when scientists and engineers discovered how to increase the capacity of integrated circuits by a factor of one million as they have in the past forty years, it enabled entrepreneurs to replace tape recorders with iPods, maps with GPS, pay phones with cell phones, two-dimensional X-rays with three-dimensional CT scans, paperbacks with electronic books, slide rules with computers, and much, much more.[3] Further, the pace of creation of new knowledge appears by almost all measures to be accelerating.[4]

Importantly, *leverage* is at work here. It is not simply the scientist, engineer and entrepreneur who benefit from progress in the laboratory or design center; it is also the factory worker who builds items such as those cited above, the advertiser who promotes them, the truck driver who delivers them, the salesperson who sells them, and the maintenance person who repairs them—not to mention the benefits realized by the user. Further, each job directly created in the chain of manufacturing activity generates, on average, another 2.5 jobs in such unrelated endeavors as operating restaurants, grocery stores, barber shops, filling stations and banks.[5] Progress enabling products such as those mentioned above in the information fields is built upon the work of a few individuals who decades ago were investigating something called solid state physics—none of whom probably ever thought about CT scans, GPS or iPods—the latter of which can enable one to hold 160,000 books in one's pocket—any more than one today can predict the breakthroughs a half century hence.[6]

---

[2] National Science Board, *Science and Engineering Indicators 2010.* Arlington, VA: National Science Foundation (NSB 10-01), Figure 3-3.

[3] In 1971, the Intel 4004 Processor had 2300 transistors. See: *http://download.intel.com/pressroom/ kits/events/moores_law_40th/MLTimeline.pdf.* In 2009, Intel released the Xeon® 'Nehalem-EX' Processor with 2.3 billion transistors. See: *http://www.intel.com/pressroom/archive/releases/2009/20090526comp. htm.*

[4] Beyond Discovery: The Path from Discovery to Human Benefit is a series of articles that explore the origins of various technological and medical advances (*www.beyonddiscovery.org/*).

[5] J. Bivens, Updated Employment Multipliers for the U.S. Economy (2003), Economic Policy Institute Working Paper, August 2003. Available at: *http://www.epi.org/page/-/old/workingpapers/epi_wp_268. pdf.*

[6] For a 64 gigabyte iPod, holding books with an average file size of 400 kilobytes.

The *Gathering Storm* report assessed America's position with respect to each of the principal ingredients of innovation and competitiveness—Knowledge Capital, Human Capital and the existence of a creative "Ecosystem." Numerous significant findings resulted—for example, with regard to Knowledge Capital it was noted that federal government funding of R&D as a fraction of GDP has *declined* by 60 percent in 40 years.[7] With regard to Human Capital, it was observed that over two-thirds of the engineers who receive PhD's from United States universities are not United States citizens.[8] And with regard to the Creative Ecosystem it was found that United States firms spend over twice as much on litigation as on research.[9] However, the most pervasive concern was considered to be the state of United States K-12 education, which on average is a laggard among industrial economies—while costing more per student than any other OECD country.[10]

So where *does* America stand relative to its position of five years ago when the *Gathering Storm* report was prepared? The unanimous view of the committee members participating in the preparation of this report is that our nation's outlook has worsened. While progress has been made in certain areas—for example, launching the Advanced Research Projects Agency-Energy—the latitude to fix the problems being confronted has been severely diminished by the growth of the national debt over this period from $8 trillion to $13 trillion.[11]

Further, in spite of sometimes heroic efforts and occasional very bright spots, our overall public school system—or more accurately 14,000 systems—has shown little sign of improvement, particularly in mathematics and science.[12] Finally, many other nations *have* been markedly progressing, thereby affecting America's relative ability to compete effectively for new factories, research laboratories, administrative

---

[7] Federal R&D was 1.92 percent of GDP in 1964 and 0.76 percent of GDP in 2004. See: *http://www. nsf.gov/statistics/nsf10314/pdf/tab13.pdf.*

[8] National Science Foundation, Division of Science Resources Statistics, *Survey of Earned Doctorates.* See *http://www.nsf.gov/statistics/nsf09311/pdf/tab3.pdf.*

[9] NSB, 2010, Appendix Tables 4-8 and 4-9; Towers Perrin, *2009 Update on U.S. Tort Cost Trends,* Appendixes 1-5.

[10] NSB, 2010, Appendix Tables 1-9, 1-10, and 1-11; and Organization for Economic Cooperation and Development, *Education at a Glance 2009: OECD Indicators;* Table B-1. See: *http://www.oecd. org/document/24/0,3343,en_2649_39263238_43586328_1_1_1_37455,00.html.*

[11] See Table 7.1, Federal Debt at the End of the Year: 1940:2015 at: *http://www.whitehouse.gov/omb/ budget/Historicals/* (accessed August 23, 2010).

[12] National Center for Education Statistics, Numbers and Types of Public Elementary and Secondary Local Education Agencies, From the Common Core of Data: School Year 2007–08. See: http://nces. ed.gov/pubs2010/2010306.pdf (accessed August 23, 2010).

centers—and *jobs*. While this progress by other nations is to be both encouraged and welcomed, so too is the notion that Americans wish to continue to be among those peoples who do prosper.

The only promising avenue for achieving this latter outcome, in the view of the *Gathering Storm* committee and many others, is through *innovation*. Fortunately, this nation has in the past demonstrated considerable prowess in this regard. Unfortunately, it has increasingly placed shackles on that prowess such that, if not relieved, the nation's ability to provide financially and personally rewarding jobs for its own citizens can be expected to decline at an accelerating pace. The recommendations made five years ago, the highest priority of which was strengthening the public school system and investing in basic scientific research, appears to be as appropriate today as then.

The *Gathering Storm* Committee's overall conclusion is that in spite of the efforts of both those in government and the private sector, the outlook for America to compete for quality jobs has further deteriorated over the past five years.

The *Gathering Storm* increasingly appears to be a Category 5.

A Few Factoids

Thirty years ago, ten percent of California's general fund went to higher education and three percent to prisons. Today, nearly eleven percent goes to prisons and eight percent to higher education.[1]

China is now second in the world in its publication of biomedical research articles, having recently surpassed Japan, the United Kingdom, Germany, Italy, France, Canada and Spain.[2]

The United States now ranks 22nd among the world's nations in the density of broadband Internet penetration and 72nd in the density of mobile telephony subscriptions.[3]

In 2009, 51 percent of *United States* patents were awarded to non-United States companies.[4]

The World Economic Forum ranks the United States 48th in quality of mathematics and science education.[5]

Of Wal-Mart's 6,000 suppliers, 5,000 are in China.[6]

There are sixteen energy companies in the world with larger reserves than the largest United States company.[7]

IBM's once promising PC business is now owned by a Chinese company.[8]

The legendary Bell Laboratories is now owned by a French company.[9]

Hon Hai Precision Industry Co. (computer manufacturing) employs more people than the worldwide employment of Apple, Dell, Microsoft, Intel and Sony combined.[10]

No new nuclear plants and no new petroleum refineries have been built in the United States in a third of a century, a period characterized by intermittent energy-related crises.[11]

Only four of the top ten companies receiving United States patents last year were United States companies.[12]

United States consumers spend significantly more on potato chips than the government devotes to energy R&D.[13]

The world's largest airport is now in China.[14]

In 2000 the number of foreign students studying the physical sciences and engineering in United States graduate schools for the first time surpassed the number of United States students.[15]

Federal funding of research in the physical sciences as a fraction of GDP fell by 54 percent in the 25 years after 1970. The decline in engineering funding was 51 percent.[16]

GE has now located the majority of its R&D personnel outside the United States.[17]

Manufacturing employment in the U.S. computer industry is now lower than when the first personal computer was built in 1975.[18]

In the 2009 rankings of the Information Technology and Innovation Foundation the U.S. was in sixth place in global innovation-based competitiveness, but ranked 40th in the rate of change over the past decade.[19]

China has now replaced the United States as the world's number one *high-technology* exporter.[20]

In 1998 China produced about 20,000 research articles, but by 2006 the output had reached 83,000 . . . overtaking Japan, Germany and the U.K.[21]

Eight of the ten global companies with the largest R&D budgets have established R&D facilities in China, India or both.[22]

During a recent period during which two high-rise buildings were constructed in Los Angeles, over 5,000 were built in Shanghai.[23]

In a survey of global firms planning to build new R&D facilities, 77 percent say they will build in China or India.[24]

China has a $196 billion positive trade balance. The United States' balance is negative $379 billion.[25]

Sixty-nine percent of United States public school students in fifth through eighth grade are taught mathematics by a teacher without a degree or certificate in mathematics.[26]

Ninety-three percent of United States public school students in fifth through eighth grade are taught the physical sciences by a teacher without a degree or certificate in the physical sciences.[27]

Of the Big Three American automakers, one is now owned by a firm in Italy (after having been previously sold by a German firm), and another is 60 percent owned by the United States government.[28]

The United States ranks 27th among developed nations in the proportion of college students receiving undergraduate degrees in science or engineering.[29]

Forty-nine percent of United States adults do not know how long it takes for the Earth to revolve around the Sun.[30]

The United States graduates more visual arts and performing arts majors than engineers.[31]

The total *annual* federal investment in research in mathematics, the physical sciences and engineering is now equal to the *increase* in United States healthcare costs every nine weeks.[32]

Bethlehem Steel marked its 100th birthday by declaring bankruptcy.[33]

The United States ranks 20th in high school completion rate among industrialized nations and 16th in college completion rate.[34]

In less than 15 years, China has moved from 14th place to second place in published research articles (behind the United States).[35]

China's real annual GDP growth over the past thirty years has been 10 percent.[36]

According to OECD data the United States ranks 24th among thirty wealthy countries in life expectancy at birth.[37]

For the next 5-7 years the United States, due to budget limitations, will only be able to send astronauts to the Space Station by purchasing rides on Russian rockets.[38]

The average American K-12 student spends four hours a day in front of a TV.[39]

China's Tsinghua and Peking Universities are the two largest suppliers of students who receive PhD's—in the United States.[40]

Sixty-eight percent of U.S. state prison inmates are high school drop-outs or otherwise did not qualify for a diploma.[41]

The United States has fallen from first to eleventh place in the OECD in the fraction 25-34 year olds that has graduated high school. The older portion of the U.S. workforce ranks first among OECD populations of the same age.[42]

When MIT put its course materials on the worldwide web, over half of the users were outside the United States.[43]

Six of the ten best-selling vehicles in the United States are now foreign models.[44]

Since 1995 the United States share of world shipments of photovolta-ics has fallen from over 40 percent to well under 10 percent—while the overall market has grown by nearly a factor of one hundred.[45]

Among manufacturers of photovoltaics, wind turbines and advanced batteries, the top ten global firms by market capitalization include two, one and one United States firms, respectively. The other firms are from China, Denmark, France, Germany, India, Spain, Taiwan and the U.K.[46]

An American company recently opened the world's largest private solar R&D facility . . . in Xian, China.[47]

By 2008, public spending in the United States on energy R&D had declined to less than half what it was three decades ago in real pur-chasing power. By 2005, private investment had declined to less than one-third of the total.[48]

A single Japanese automobile model constitutes about half of the U.S. hybrid market.[49]

Last year Mitsubishi introduced the world's first mass-produced all-electric car.[50]

A Japanese company produces over 75 percent of the world's nickel-metal hydride batteries used in vehicles.[51]

Japan has 1524 miles of high speed rail; France has 1163; and China just passed 742 miles. The United States has 225. China has 5612 miles now under construction and one plant produces 200 trains each year capable of operating at 217 mph. The United States has none under construction.[52]

Roughly half of America's outstanding public debt is now foreign-owned—with China the largest holder.[53]

The increase in cost of higher education in America has substantially surpassed the growth in family income in recent decades. United States current and former students have amassed $633 billion in student loan debt.[54]

There are 60 new nuclear power plants currently being built in the world. One of these is in the United States.[55]

In 2008, 770,000 people worked in the United States correction sector, a number which is projected to grow. During the same year there were 880,000 workers in the entire United States automobile manufacturing sector.[56]

Between 1996 and 1999, 157 new drugs were approved in the United States. In a corresponding period ten years later the number dropped to 74.[57]

All the National Academies *Gathering Storm* committee's recommendations could have been fully implemented with the sum America spends on cigarettes each year—with $60 billion left over.[58]

Youths between the ages of 8 and 18 average seven-and-a-half hours a day in front of video games, television and computers—often multi-tasking.[59]

In 2007 China became second only to the United States in the estimated number of people engaged in scientific and engineering research and development.[60]

In January 2010, China's BGI made the biggest purchase of genome sequencing equipment ever.[61]

In May 2010, a supercomputer produced in China was ranked the world's second-fastest.[62]

Almost one-third of U.S. manufacturing companies responding to a recent survey say they are suffering from some level of skills shortages.[63]

According to the ACT College Readiness report, 78 percent of high school graduates did not meet the readiness benchmark levels for one or more entry-level college courses in mathematics, science, reading and English.[64]

ENDNOTES

1 J. Steinhauer, Schwarzenegger Seeks Shift From Prisons to Schools, *The New York Times*, January 6, 2010.

2 J. Karlberg. Biomedical Publication Trends by Geographic Area. *Clinical Trial Magnifier*. 2 (12), December, 2009.

3 S. Dutta and I. Mia, *Global Information Technology Report 2009–2010: ICT for Sustainability*, World Economic Forum, 2010.

4 T. Donohue, Testimony to the House Committee on Science and Technology on The Reauthorization of the America COMPETES Act, January 20, 2010. Available at: *https://www.uschamber.com/issues/testimony/2010/100119_americacompetes.htm*. See also: *http://www.ificlaims.com/IFI%202009%20patents%20011210%20final.htm*.

5 World Economic Forum, *The Global Competitiveness Report 2009-2010*, Available at: *http://www.weforum.org/pdf/GCR09/Report/Countries/United%20States.pdf*.

6 P. Goodman and P. Pan, Chinese Workers Pay for Wal-Mart's Low Prices: Retailer Squeezes Its Asian Suppliers to Cut Costs, *The Washington Post*, February 8, 2004.

7 See: *http://www.petrostrategies.org/Links/Worlds_Largest_Oil_and_Gas_Companies_Sites.htm* (accessed August 23, 2010).

8 N. Augustine, *Is America Falling Off the Flat Earth?* National Academy of Sciences, National Academy of Engineering, Institute of Medicine, 2007, page 17; Available at: *http://www.nap.edu/openbook.php*.

9 J. Zarroli, French Telecom Company Alcatel Merging with Lucent, *NPR*, April 3, 2006.

10 J. DiPietro, Silicon Valley Is Dead, The Motley Fool, July, 27, 2010. Available at: *http://www.fool.com/investing/general/2010/07/27/silicon-valley-is-dead.aspx*.

11 N. Augustine, 2007.

12 T. Donohue, 2010.

13 For 2009 U.S. potato chip sales of $7.1 billion, see *https://www.aibonline. org/resources/statistics/2009snack.htm*. For U.S. federal government spending on energy R&D of $5.1 billion, see American Energy Innovation Council, *A Business Plan for America's Energy Future*, 2010.

14 Beijing's Giant Airport Terminal To Open, *BusinessWeek*, February 27, 2008.

15 *Measuring the Moment: Innovation, National Security, and Economic Competitiveness*, The Task Force on the Future of American Innovation. Available at: *http://futureofinnovation.org/PDF/BII-FINAL-HighRes-11-14-06_nocover.pdf*.

16 N. Augustine, 2007.

17 R. Hira, U.S. Workers in a Global Job Market, *Issues in Science and Technology*, Spring 2009, Available at: *http://www.issues.org/25.3/hira.html*.

18 A. Grove, How to Make an American Job Before It's Too Late, *Bloomberg*, July 1, 2010.

19 Information Technology and Innovation Foundation, *The Atlantic Century: Benchmarking EU & U.S. Innovation and Competitiveness*, February 2009. See: *http://www.itif.org/files/2009-atlantic-century.pdf*.

20 T. Meri, *Eurostat: Statistics in Focus*, 2009; Available at: *http://epp.eurostat. ec.europa.eu/cache/ITY_OFFPUB/KS-SF-09-025/EN/KS-SF-09-025-EN.PDF*.

21 J. Adams, Science heads east, *New Scientist*, Volume 205, Issue 2742, January 9, 2010.

22 From R. Atkinson, The Globalization of R&D and Innovation: How Do Companies Choose Where to Build R&D Facilities? Testimony, Committee on Science and Technology, Subcommittee on Technology and Innovation, U.S. House of Representatives, October 4, 2007.

23 H. Fineberg, Address to The Academy of Medicine, Engineering & Science of Texas Annual Meeting, January 5, 2006.

24 T. Goldbrunner, Y. Doz, and K. Wilson The Well-Designed Global R&D Network, *Strategy+Business*, May 30, 2006.

25 For China, J.R. Wu, China's Exports Turn Upward in December, *The Wall Street Journal*, January 11, 2010; for the United States, see: *http://www.census.gov/ foreign-trade/Press-Release/2010pr/01/ft900.pdf*.

26 National Center for Education Statistics, *Qualifications of the Public School Teacher Workforce: Prevalence of Out-of-Field Teaching 1987-1988 and 1999-2000*, Washington, DC: U.S. Department of Education, 2003.

27 Ibid.

28 D. Silver, General Motors Files Bankruptcy, WStreet.com, June 2, 2009. Available at: *http://www.wstreet.com/investing/stocks/17551_general_motors_files_ bankruptcy.html*.

29 Organization for Economic Cooperation and Development, *Education at a Glance 2009: OECD Indicators*; Table A-3.5.

30 National Science Board, *Science and Engineering Indicators: 2010*, Arlington, VA, Appendix Table 7-9.

31 National Center for Education Statistics, *Digest of Education Statistics: 2009*, Washington, DC. See: *http://www.nces.ed.gov/programs/digest/d09/tables/dt09_271. asp?referrer=list*.

32 For figures on research, see NSB, 2010, Appendix Table 4-23. For figures on healthcare spending see Centers for Medicare & Medicaid Services, National Health Expenditures Aggregate, Per Capita Amounts, Percent Distribution, and Average Annual Percent Growth, by Source of Funds: Selected Calendar Years 1960-2008. Available at: *http://www.cms.gov/NationalHealthExpendData/downloads/tables.pdf.*

33 N. Augustine, Learning to Compete, *Princeton Alumni Weekly*, March 7, 2007. Available at: *http://www.princeton.edu/~paw/archive_new/PAW06-07/09-0307/perspective.html.*

34 OECD, 2009. Rankings include OECD members and partners, and college graduation ranking is based on Tertiary-A institutions. See: Tables A2.1 and A3.1 in *http://www.oecd.org/document/24/0,3343,en_2649_39263238_43586328_1_1_1_1,00.html.*

35 J. Pomfret, China pushing the envelope on science, and sometimes ethics, *The Washington Post*, June 28, 2010.

36 International Monetary Fund data available here: *http://www.imf.org/external/pubs/ft/weo/ 2010/01/weodata/weoselgr.aspx.*

37 Organization for Economic Cooperation and Development, *Health at a Glance 2009*, Paris, 2009. Available at: Available at*: http://www.oecd.org/health/healthataglance.*

38 Obama aims to send astronauts to Mars orbit in 2030s, PhysOrg.com, April 15, 2010. Available at: *http://www.physorg.com/news190564316.html.*

39 P. McDonough, TV Viewing Among Kids at an Eight-Year High, *Nielsen Wire*, October 26, 2009. Available at: *http://blog.nielsen.com/nielsenwire/media_entertainment/tv-viewing-among-kids-at-an-eight-year-high/.*

40 J. Mervis, Top Ph.D. Feeder Schools Are Now Chinese, *Science*, July 11, 2008.

41 C. Harlow, Educational and Correctional Populations, *Bureau of Justice Statistics Special Report*, January 2003. Available at*: http://bjs.ojp.usdoj.gov/content/pub/pdf/ecp.pdf.*

42 OECD, 2009. See Chart A1.2 at *http://www.oecd.org/dataoecd/41/25/43636332.pdf.*

43 MIT OpenCourseWare, *2005 Program Evaluation Findings Report*, June 5, 2006. Available at: *http://ocw.mit.edu/ans7870/global/05_Prog_Eval_Report_Final.pdf.*

44 The Best Selling Cars of 2009, *U.S. News and World Report*, January 4, 2010. Available at: *http://usnews.rankingsandreviews.com/cars-trucks/daily-news/100104-The-Best-Selling-Cars-Of-2009/.*

45 From the Statement of Arun Majumdar, Director, Advanced Research Projects Agency-Energy (ARPA-E), U.S. Department of Energy, Before the Committee on Science and Technology, U.S. House of Representatives, January 27, 2010.

46 The President's Economic Recovery Advisory Board, Memorandum for the President on Energy, the Environment and Technology, June 17, 2009. Available at: *http://www.whitehouse.gov/sites/default/files/microsites/090520_perab_climateMemo.pdf.*

47 Applied Materials Opens Solar Technology Center in Xian, China, *TechOn*, October 27, 2009. Available at: *http://techon.nikkeibp.co.jp/english/NEWS_EN/20091027/176977/.*

48 J. Dooley, U.S. Federal Investments in Energy R&D: 1961-2008, U.S. Department of Energy, PNNL-17952, October 2008. Available at: *http://www.green techhistory.com/wp-content/uploads/2009/07/federal-investment-in-energy-rd-2008. pdf*. See also: D. Kammen and G. Nemet, Reversing the Incredible Shrinking Energy R&D Budget, *Issues in Science and Technology*, Fall 2005.

49 Best Selling Hybrid Cars in 2009, *Hybrid Cars*, March 4, 2010. Available at: *http://www.hybrid-cars.org/announcements/top-hybrid-car-2009*.

50 Mitsubishi unveils first mass-market electric car from a major car maker, *The Guardian*, January 20, 2010. See: *http://www.guardian.co.uk/technology/2009/ jan/20/ greentech-travelandtransport*.

51 M. Huckerbee, S. Wong, T. Cai, China's intervention in global M&A heats up while AML private actions cool down, Mallesons Stephen Jaques website, November 2009, See: *http://www.mallesons.com/publications/2009/Nov/10131333w.htm*.

52 K. Richburg, China is pulling ahead in worldwide race for high-speed rail transportation, *The Washington Post*, May 12, 2010.

53 Information on foreign holdings of U.S, treasury securities is available at: *http://www.ustreas.gov/tic/mfh.txt*. Total debt held by the public is available at: *http:// www.treasurydirect.gov/NP/NPGateway*.

54 A. Damast, Asking for Student Loan Forgiveness, *BusinessWeek*, March 24, 2009.

55 International Atomic Energy Agency. See: *http://www.iaea.org/cgi-bin/ db.page.pl/pris.reaucct.htm* (Accessed August 26, 2010).

56 S. Kirchhoff, *Economic Impacts of Prison Growth*, Congressional Research Service, 7-5700, April 13, 2010

57 Approvals of new molecular entities (NMEs) are counted. See J. Owens, 2006 Drug Approvals: Finding the Niche, Nature Reviews Drug Discovery, February 2007. Available at: *http://www.nature.com/nrd/journal/v6/n2/fig_tab/ nrd2247_F1.html*; M. Martino, 2007 FDA Approvals, *Fierce Biotech*, January 14, 2008. Available at: *http://www.fiercebiotech.com/special-reports/2007-fda-approvals*; and M. Arnold, FDA BLA approvals rose in 2009 while NMEs stumbled, *Medical Marketing and Media*, December 31, 2009. Available at: *http://www.mmm-online. com/fda-bla-approvals-rose-in-2009-while-nmes-stumbled/article/160496/*.

58 For spending on cigarettes, see: *http://www.cdc.gov/tobacco/data_statistics/ fact_sheets/economics/econ_facts/*. For cost estimates of the *Gathering Storm* committee's recommendations, see National Academy of Sciences, National Academy of Engineering, and Institute of Medicine, *Rising Above the Gathering Storm: Energizing and Employing America for a Brighter Economic Future*, Washington, DC, 2007, Appendix E.

59 V. Rideout, U. G. Foehr, and D. F. Roberts, Generation M^2: Media in the Lives of 8- to 18-Year-Olds, Kaiser Family Foundation, January 2010.

60 NSB, 2010.

61 J. Pomfret, China pushing the envelope on science, and sometimes ethics, *Washington Post*, June 28, 2010.

62 Ibid.

63 Deloitte, Oracle, and the Manufacturing Institute, *People and profitability: A time for change*, 2009.

64 Note that the ACT estimates that students meeting the readiness standard in a given subject have a 75 percent chance of getting a C and a 50 percent chance of getting a B in an entry level course. Information available at the ACT website: *http://www.act.org/news/releases/2008/crr.html.*

1.0
The Gathering Storm, Revisited

"It is amazing the insights one can get sitting in an airport waiting for a flight. Last week I found myself at Heathrow airport in London sitting with two businessmen. The stories they told were remarkable.

One businessman was on his way home from China. He had just spent a week working out the details to build a major manufacturing facility at a booming Chinese industrial park. The development authority promised expedited zoning procedures to facilitate rapid construction. Another government entity offered to line up job fairs to recruit workers. The mayor said they will provide free transportation services from downtown to the factory for employees for two years. The local university promised an intern program for engineering students. And on it went.

The other businessman relayed the challenges he faced with his factory in America. He wanted to double the size of the factory, but was informed he needed a new environmental impact study before he could approach zoning authorities. He sought a meeting with federal officials who were managing "stimulus" funds that he hoped would help finance the expansion. But he was informed he could not meet with the assistant secretary because it would possibly suggest a 'conflict of interest'."[a]

John Hamre,
Former Deputy Secretary of Defense

---

[a]Communication from John Hamre, July 27, 2010.

THE ASSIGNMENT

In 2005 the National Academies was requested, on a bipartisan basis, to conduct a review of America's competitiveness in the rapidly evolving global marketplace and, as appropriate, to offer specific actions that could be taken by federal policymakers to ensure the nation's position as a prosperous member of the global economy of the twenty-first century. Ten weeks were allotted for the conduct of the study, after which a 500-page volume was produced bearing the title, *"Rising Above the Gathering Storm: Energizing and Employing America for a Brighter Economic Future."*

The initial request for the study was made jointly by Senator Lamar Alexander and Senator Jeff Bingaman, both of the Senate Committee on Energy and Natural Resources, and was endorsed by Representative Sherwood Boehlert and Representative Bart Gordon of the House Committee on Science and Technology. The study's conduct was supported by numerous members of the House of Representatives and Senate from both parties as well as members of the Administration, including the President.

In responding to the above request, the National Academies convened a twenty-person committee composed of individuals having highly diverse professional backgrounds. These included chief executive officers of major corporations, presidents of public and private universities, scientists and engineers (including three Nobel Laureates), former presidential appointees, and the superintendent of a state school system. The committee benefited greatly from the inputs of well over one hundred experts in specific fields and the support of a remarkable staff directed by Dr. Deborah Stine. Before being approved the report was subjected to an anonymous critique by 37 members of the National Academies.

REVISITING THE GATHERING STORM

The present document was reviewed and endorsed by members of the same committee that had prepared the initial *Gathering Storm* report, with the exceptions of Dr. Robert Gates and Dr. Steven Chu, who are now serving as Secretary of Defense and Secretary of Energy, respectively. Further, the members of the committee observe with deep regret the death in 2008 of our much respected colleague and friend, Nobel Laureate Joshua Lederberg.

The original *Gathering Storm* report concluded that the fundamental measure of competitiveness is *quality jobs*. It is *jobs* that to a considerable degree define the quality of life of a nation's individual citizens. Further, it is the tax revenues derived from the earnings of those individual citizens and the firms that employ them that make possible the benefits widely expected from government—including, but by no means limited to, healthcare, national security, physical infrastructure, education and unemployment assistance.

Substantial evidence continues to indicate that over the long term the great majority of newly created jobs are the indirect or direct result of advancements in science and technology, thus making these and related disciplines assume what might be described as disproportionate importance. A variety of economic studies over the years reveals that half or more of the growth in the nation's Gross Domestic Product (GDP) in recent decades has been attributable to progress in technological innovation.[1]

Advancements in these fields have led not only to the creation of large numbers of quality jobs, they have also made it possible for hundreds of millions of people around the globe to compete with Americans for these same jobs. In particular, the advent of modern aircraft and modern information systems has made it feasible to move *objects,* including humans, around the world at nearly the speed of sound—and *ideas* at the speed of light, the latter with little regard for geopolitical borders. In describing the consequences of this development, writer Tom Friedman notes that "Globalization has accidentally made Beijing, Bangalore and Bethesda next-door neighbors."

Whereas in the past, citizens of any one nation generally had to compete for jobs with their neighbors living in the same community, in the future they will increasingly be required to compete with individuals who live half-way around the world.

Software written in India is now shipped to the United States in milliseconds to be integrated into systems that same day. Flowers grown in Holland are flown overnight for sale in New York the next morning. Magnetic Resonance Images (MRI's) of patients in United States hospitals are read moments later by radiologists in Australia. Pilots stationed in the United States guide unmanned aircraft to attack targets in Afghanistan. United States accounting firms prepare United States citizens income taxes using accountants located in Costa Rica and Switzerland. Water collected in France is sold in grocery stores in California. The receptionist in an office in Washington, DC lives in Pakistan. A physician in New York removes the gall bladder of a patient in France with the help of a remotely controlled robot. And so it goes in a world where, as described by Frances Cairncross writing in *The Economist,* "Distance is Dead."

---

[1] M. J. Boskin and L. J. Lau. Capital, Technology, and Economic Growth. In N. Rosenberg, R. Landau, and D. C. Mowery, eds. *Technology and the Wealth of Nations.* Stanford, CA: Stanford University Press, 1992.

If Americans are to compete for quality jobs in such a world—one where three *billion* new would-be capitalists entered the job market upon the restructuring of many of the world's political systems late in the last century—they will need help from their government . . . at all levels . . . as well as from themselves. The latter includes preparing for the growing educational demands of quality jobs and continuing to *maintain* their skills in a circumstance where the half-life of new technical knowledge may be measured in terms of a few years or, in some cases, even a few months.

The *Gathering Storm* committee contends that it is strongly in America's interest for all nations to prosper. Aside from its humanistic merit this outcome should produce a safer world for everyone, one with more products for United States consumers and more consumers for United States products. But the committee also expressed its commitment to helping America to be among those nations enjoying the fruits of what it hopes will be truly global prosperity. In the latter regard, the committee concluded that *the United States appears to be on a course that will lead to a declining, not growing, standard of living for our children and grandchildren.*

The likelihood of a more promising outcome can be enhanced by implementing the four overarching recommendations (via twenty specific actions) offered in the original *Gathering Storm* report . . . and to sustain the effort needed to reach fruition. It is noteworthy that America's current predicament was not generated in a decade, nor will it be resolved in a decade. Staying-power is essential.

Recommendations

The *Gathering Storm* committee offered four overarching recommendations. These are highly interdependent. For example, to produce more researchers but not increase research spending would be highly counterproductive. In order of assigned importance, the four recommendations can be summarized as follows:

I.   Move the United States K-12 education system in science and mathematics to a leading position *by global standards.*

II. Double the real federal investment in basic research in mathematics, the physical sciences, and engineering over the next seven years (while, *at a minimum,* maintaining the recently doubled *real* spending levels in the biosciences).

III. Encourage more United States citizens to pursue careers in mathematics, science, and engineering.

IV. Rebuild the competitive ecosystem by introducing reforms in the nation's tax, patent, immigration and litigation policies.

Implementing Actions

In support of the above general recommendations, the National Academies offered 20 specific implementing actions[2]:

❖ **"10,000 Teachers Educating 10 Million Minds" (focuses on K-12 education, the committee's unanimous highest priority).**

- **Provide 10,000 *new* mathematics and science teachers each year by funding competitively awarded 4-year scholarships for U.S. citizens at U.S. institutions that offer special programs leading to core degrees in mathematics, science, or engineering *accompanied by a teaching certificate*. On graduation, participants would be required to teach in a public school for five years and, one hopes, beyond that time by choice.**

- **Strengthen the skills of 250,000 *current* teachers by such actions as subsidizing the achievement of master's degrees (in science, mathematics, or engineering) and participation in workshops, and create a world-class mathematics and science curriculum available for voluntary adoption by local school districts throughout the nation.**

- **Increase the number of teachers qualified to teach Advanced Placement courses and the number of students enrolled in those courses by offering financial bonuses both to high-performing teachers and to students who excel.**

---

[2] See National Academy of Sciences, National Academy of Engineering, Institute of Medicine, *Rising Above the Gathering Storm: Energizing and Employing America for a Brighter Economic Future,* Washington, DC: National Academies Press, 2007, for a discussion of the recommendations.

❖ "Sowing the Seeds" (focuses on funding for research).

- Increase federal basic-research funding in the physical sciences, mathematics, and engineering by a real ten percent each year over the next seven years.

- Provide research grants each year to 200 early-career researchers, payable over five years.

- Provide an incremental $500 million per year for at least five years to modernize the nation's aging research facilities, with the expenditures overseen by a National Coordination Office for Research Infrastructure to be in the White House Office of Science and Technology Policy.

- Allocate eight percent of government research funds to pursuits specifically chosen at the discretion of local researchers and their managers, with emphasis on projects potentially offering a high payoff even though accompanied by substantial risk.

- Establish an ARPA-E in the Department of Energy patterned after the highly successful DARPA in the Department of Defense but focused on major breakthroughs in energy security.

- Institute a Presidential Innovation Award to stimulate advances serving the national interest.

❖ "Best and Brightest" (focuses on higher education).

- Provide 25,000 competitively awarded undergraduate scholarships each year of up to $20,000 per year for 4 years in the physical and life sciences, mathematics, and engineering for U.S. citizens attending U.S. institutions.

- Provide 5,000 competitively awarded portable graduate fellowships each year of up to $20,000 per year in fields of national need.

- Grant tax credits to employers that support continuing education for practicing scientists and engineers.

- Continue to improve visa processing for international students.

- Offer a one-year visa extension to PhD recipients in science, technology, engineering, mathematics or other fields of national need; grant automatic work permits to those meeting security requirements and obtaining employment; provide a preferential system for acquiring citizenship for those who complete their degrees; and repeal the mandatory "go-away" provision now in U.S. immigration law.

- Offer preferential visas to applicants who have special skills in mathematics, science, engineering, and selected languages.

- Modify the "deemed export" law whereby faculty currently may be required to obtain export licenses to teach a technology class that includes a foreign student even if the material covered is unclassified.

❖ "Incentives for Innovation" (focuses on the innovation environment).

- Adopt a "first-to-file" patent system and increase employment of the U.S. Patent and Trademark Office to permit accelerated handling of patent matters.

- Expand and make permanent the R&D tax credit that has been extended eleven times since it was first enacted in 1981 but never made permanent.

- Restructure the corporate income-tax laws to help make firms that create jobs in the United States more competitive.

- Increase broadband Internet access throughout the nation.

Additional Observation

The *Gathering Storm* report was prepared shortly after the nation's research budget *in the health sciences* had, over a five-year period, doubled. The *Gathering Storm* review thus focused on the physical sciences, mathematics and engineering, fields for which real funding had been stagnant for decades. However, shortly after the "doubling" in the health sciences was achieved, the funding for that activity was permitted to erode once again—the exception being a major one-time, two-year funding infusion as part of the American Recovery and Reinvestment Act.

Many of the findings of the Academies' study regarding the physical sciences, mathematics and engineering now pertain to the biological sciences as well. This is a particularly significant development given the evolving interdependency among disciplines— wherein, for example, automotive fuels and plastics are now being made using biological processes, and biocomputing (the use of biological molecules to perform computational calculations) is in the early research stage. Also, there is a critical need for physicists and mathematicians to help mine the vast seas of data coming from genome studies being done to understand the development and treatment of cancers and other diseases. During the past century life expectancy in America has increased by over 50 percent due in substantial part to advances in the health sciences—indicating the impact of various types of innovation.[3]

---

[3] NAS-NAE-IOM, 2007, p. 51.

2.0
Efforts to Avert the Storm

INITIAL FOLLOW-UP

The *Gathering Storm* report concluded that America was in substantial danger of losing its economic leadership position and suffering a concomitant decline of the standard of living of its citizens because of a looming inability to compete for jobs in the global marketplace.

In the weeks following the *Gathering Storm* report's release, over one hundred editorials and op-eds appeared in the nation's newspapers, at least one in every state, addressing the issues raised in the report. Virtually all supported the National Academies' conclusions and joined in the call for action. President George W. Bush included many of the report's recommendations in his 2006 State of the Union Address and in the days immediately following the address traveled extensively, speaking in part about the report's highest priority findings—K-12 education and basic research.

Implementing-legislation with 62 co-sponsors was promptly introduced in the United States Senate. A series of hearings was held in both the House and the Senate and the "America COMPETES Act" was forwarded to the House and Senate floor with strong bipartisan support—something the *Gathering Storm* effort has enjoyed throughout the first five years of its existence.

During 2007, the House of Representatives, in two key actions, approved the necessary authorizing legislation by votes of 389-22 and 397-20 (two votes were taken for procedural reasons). The authorizing legislation subsequently passed the Senate by a vote of 88-8. Final approval in the House of the America COMPETES authorization act was by unanimous consent following 367-57 approval of the conference report. In what perhaps might best be described as a system failure, virtually no funds to implement the *Gathering Storm* recommendations were included in the final version of the Fiscal 2008 Appropriations Act (although some 10,000 earmarks survived). The *Gathering Storm* recommendations required approximately $19 billion per year for implementation, once a transition phase was completed. Starting with the fiscal 2008 supplemental budget, funding for the relevant agencies of the Department of Energy Office of Science, the National Science Foundation and the National Institute of Standards and Technology laboratories has been on a trajectory that, if sustained, will result in a doubling by 2017. In contrast, funding of the STEM education-related recommendations lagged.

During the first two years following the release of the *Gathering Storm* report the principal impact of the efforts by the Academies and a wide array of interested constituencies—including the Council on Competitiveness, the Business Roundtable, the American Association for the Advancement of Science, the American Physical Society, the Association of American Universities, the Association of Public and Land-Grant Universities, and others—was to forestall actions that would otherwise have *diminished* America's competitiveness. A private-sector organization, the National Math and Science Initiative, was established based on the *Gathering Storm* recommendations to increase participation in Advance Placement courses in high school and to provide additional teachers qualified in mathematics and science.[1]

SUBSEQUENT FOLLOW-UP

Responding to the severe downturn of the economy in the fall of 2008, "stimulus legislation," designated the American Recovery and Reinvestment Act, or ARRA, was introduced. A special hearing chaired by Speaker Nancy Pelosi was held, in part to address the *long-term* implications of any potential legislation. During the hearing witnesses noted that the *Gathering Storm* report emphasized the need for investments for the longer term—particularly in K-12 education and university research. Legislation that was eventually approved provided funding to implement many of the report's recommendations. President Obama, who had previously endorsed improvements to the nation's K-12 education system and the addition of funds for science, including a major increase in funding at the National Institutes of Health, signed ARRA on February 17, 2009.

---

[1] See http://www.nationalmathandscience.org/.

This, together with other legislation that was enacted, increased total federal support for all aspects of K-12 education by a projected $59 billion between 2009 and 2010, provided scholarships for a number of future mathematics and science teachers and provided funding for the Advanced Research Projects Agency-Energy (ARPA-E) patterned after the Department of Defense's Defense Advanced Research Projects Agency.[2] Processing of student visas was improved, reducing the delays and uncertainties that resulted from post-9/11 changes; however, this continues to be a deterrent to many talented foreign students and professionals.

OVERALL STATUS OF FOLLOW-UP

Table 2-1, derived from an assessment conducted by the Congressional Research Service, summarizes recent Congressional actions, or lack thereof, in response to each of the National Academies' recommendations and implementing actions in *Rising Above the Gathering Storm*.

Today, the American Recovery and Reinvestment Act, intended as a one-time action, is nearing expiration. Without new actions the precipitous reduction in efforts that were being funded by that mechanism will be very damaging to America's future ability to compete for jobs in the global marketplace. Similarly, authorization for the America COMPETES Act requires renewal this year as it too is scheduled to expire.

Thus, the *Gathering Storm* effort as viewed in the middle of 2010, although still enjoying strong support in the White House and in both parties in the Congress, finds itself at a tipping point. The issue at stake is whether funding to help assure that Americans can compete for quality jobs will be provided on a *sustained* basis. The budgetary pressures now faced by the nation make such investments extremely difficult; however, if not made the future consequences in terms of unemployment and related costs will likely be immense. In the judgment of the National Academies *Gathering Storm* committee, failure to support a strong competitiveness program will have dire consequences for the nation as a whole as well as for its individual citizens.

---

[2] Regarding increased K-12 education spending, based on a total of $38.8 billion in federal K-12 spending in 2008, and a projected $137.1 billion for 2009 and 2010 combined. See: http://www.whitehouse.gov/sites/default/files/omb/budget/fy2011/assets/hist09z9.xls.

SUMMARY

The two highest priority actions for the nation, in the view of the *Gathering Storm* committee, are to provide teachers in every classroom qualified to teach the subject they teach and to double the federal investment in research—the latter of which would be competitively awarded and largely conducted by the nation's research universities as opposed to government facilities.

Overall, the steps recently taken to strengthen the nation's basic research program have been substantial—albeit tenuous because of the one-time funding mechanism employed. Some steps taken to enhance K-12 education have been noteworthy as well, but in terms of actual implementation have fallen far short of the *Gathering Storm* committee's recommendations. Similarly, such actions as increasing the granting of H-1B visas; making the R&D tax credit permanent; changing intellectual property laws; modernizing export control policies; and assuring that qualified math and science teachers are available to every student, have languished.

TABLE 2-1 **Implementation Status of Recommendations from *Rising Above the Gathering Storm***

RECOMMENDATIONS AND ACTION STEPS	CONGRESSIONAL ACTIONS
Recommendation A: Increase America's talent pool by vastly improving K–12 science and mathematics education.	
Action A-1: Annually recruit 10,000 science and mathematics teachers by awarding 4-year scholarships and thereby educating 10 million minds.	
A-1-1. Provide merit-based, 4-year scholarships of up to $20,000 per year to students who commit to 5 years of teaching after obtaining bachelor's degrees in STEM fields and concurrent certification as K-12 science and mathematics teachers.	About $225 million appropriated for Robert Noyce Teacher Scholarship program at NSF over FY 2008-2010 ($10,000 stipends to juniors/seniors).
A-1-2. Award matching grants of $1 million a year for up to 5 years for universities to establish integrated 4-year undergraduate programs leading to bachelor's degrees in STEM fields *with a teacher certification.*	Over $450 million authorized, about $3 million appropriated for Teachers for a Competitive Tomorrow program over FY 2008-2010.
Action A-2: Strengthen the skills of 250,000 teachers through training and education programs at summer institutes, in master's programs, and in Advanced Placement (AP) and International Baccalaureate (IB) training programs.	
A-2-1. Provide matching grants to 1- to 2-week summer institutes to upgrade the skills of as many as 50,000 practicing teachers each summer.	Appropriated funds not specified (no line item). Obama Administration argues that this corresponds to DOE Academies Creating Teacher Scientists program.
A-2-2. Provide grants to universities to offer, over 5 years, 50,000 current middle and high school STEM teachers part-time master's degree programs.	$375 million authorized, about $3 million appropriated over FY 2008-2010.
A-2-3. Train an additional 70,000 AP or IB and 80,000 pre-AP or pre-IB instructors to teach advanced courses in STEM fields.	$45.8 million appropriated in FY 2010 for high-need schools.
A-2-4. Convene a national panel to develop rigorous K-12 materials as a *voluntary* national curriculum.	The Department of Education established a related panel in 2008 that met and commissioned several papers on undergraduate STEM education
Action A-3: Enlarge the pipeline of students who are prepared to enter college and graduate with a degree in science, engineering, or mathematics by increasing the number of students who pass AP and IB science and mathematics courses.	

TABLE 2-1 Continued

RECOMMENDATIONS AND ACTION STEPS	CONGRESSIONAL ACTIONS
A-3-1. Increase the number of students who take AP or IB STEM courses to 1.5 million, and triple the number who pass to 700,000. Student incentives to include 50% exam fee rebates and $100 mini-scholarships for each passing score on AP/IB science or mathematics exams.	$45.8 million appropriated in FY2010 for reimbursing low-income students in high-need schools for AP/IB test fees.
Other actions: Specialty STEM high schools, summer internships for middle and high school students	About $89 million authorized, but funds not appropriated or not specified.

Recommendation B: Sustain and strengthen the nation's traditional commitment to long-term basic research that has the potential to be transformational to maintain the flow of new ideas that fuel the economy, provide security, and enhance the quality of life.

Action B-1: Increase the federal investment in long-term basic research by 10% each year over the next 7 years, with special attention to the physical sciences, engineering, mathematics, and information sciences and to Department of Defense (DOD) basic-research funding.	Authorization and appropriation levels over FY 2008-2010 largely reflect this recommendation for NSF, NIST, and DOE Office of Science.
Action B-2: Provide new research grants of $500,000 each annually, payable over 5 years, to 200 of the nation's most outstanding early-career researchers.	About $75 million authorized for DOE, appropriated funds not specified; about $550 million authorized, $700 million appropriated to NSF in the FY2009 Omnibus and ARRA.
Action B-3: Institute a National Coordination Office for Advanced Research Instrumentation and Facilities to manage a fund of $500 million in incremental funds per year over the next 5 years.	OSTP states that ARRA provides funding for research infrastructure that addresses some of these concerns. OSTP has also been directed in legislation to identify deficiencies in federal research facilities and coordinate responses.
Action B-4: Allocate at least 8% of the budgets of federal research agencies to discretionary funding of high-risk, high-payoff research.	America COMPETES contains "sense of the Congress" language encouraging agencies to allocate a portion of basic research funding to transformative projects.
Action B-5: Create an Advanced Research Projects Agency-Energy (ARPA-E) with initial annual funding of $300 million, increasing to $1 billion over 5-6 years.	$830 million appropriated in the FY2009 Omnibus and ARRA.
Action B-6: Institute a Presidential Innovation Award to stimulate scientific and engineering advances in the national interest.	The existing "National Medal of Technology" has been renamed the "National Mecal of Technology and Innovation"

TABLE 2-1 Continued

RECOMMENDATIONS AND ACTION STEPS	CONGRESSIONAL ACTIONS
Recommendation C: Make the United States the most attractive setting in which to study and perform research so that we can develop, recruit, and retain the best and brightest students, scientists, and engineers from within the United States and throughout the world.	
Action C-1: Increase the number and proportion of U.S. citizens who earn bachelor's degrees in STEM fields by providing 25,000 new 4-year competitive undergraduate scholarships each year to be distributed to states on the basis of the size of their congressional delegations and awarded via national examinations.	About $350 million authorized and $130 million appropriated to NSF over FY2009-2010 for related programs, such as the STEM Talent Expansion Program and the Research Experience for Undergraduates Program.
Action C-2: Increase the number of U.S. citizens pursuing graduate study in "areas of national need" by funding 5,000 new graduate fellowships each year through NSF, with annual stipend levels of $30,000, plus $20,000 for tuition and fees.	About $640 million authorized and $475 million appropriated over FY 2009-2010 (including ARRA) to NSF for existing programs such as Graduate Research Fellowships, Integrative Graduate Education and Research Traineeships, and Protecting America's Competitive Edge Graduate Fellowships.
Action C-3: Provide a federal tax credit to encourage employers to make continuing education available (either internally or through colleges and universities) to practicing scientists and engineers.	Not acted on.
Action C-4: Continue to improve visa processing for international students and scholars.	Not addressed by Congress, but various sources report that the Department of State has made significant progress in streamlining security screening.
Action C-5: Provide a 1-year automatic visa extension to international students who receive doctorates or the equivalent in STEM fields at qualified U.S. institutions, and provide them with automatic work permits if they are offered employment by a U.S.-based employer and pass a security screening test.	By regulation, the Office of Citizenship and Immigration Services extended the optional practical training period for F-1 nonimmigrant students with STEM degrees from 12 to 29 months, and amended regulations to allow for automatic extensions of status and employment authorizations for F-1 students with pending H-1B applications.
Action C-6: Institute a new skills-based, preferential immigration option giving persons with doctoral-level education and science and engineering skills priority in obtaining U.S. citizenship. Increase the number of H1-B visas for applicants with doctorates from U.S. universities by 10,000.	Prior legislation exempts up to 20,000 aliens holding a master's or higher degree from the annual cap on H1-B visas.

EFFORTS TO AVERT THE STORM

TABLE 2-1 Continued

RECOMMENDATIONS AND ACTION STEPS	CONGRESSIONAL ACTIONS
Action C-7: Reform the current system of "deemed exports," giving international students and researchers engaged in fundamental research in the United States with access to United States research equipment and information that is comparable to that provided to U.S. citizens.	Not acted on.

Recommendation D: Ensure that the United States is the premier place in the world to innovate; invest in downstream activities such as manufacturing and marketing; and create high-paying jobs based on innovation by such actions as modernizing the patent system, realigning tax policies to encourage innovation, and ensuring affordable broadband access.

Action D-1: Enhance intellectual-property protection for the 21st-century global economy to ensure that systems for protecting patents and other forms of intellectual property underlie the emerging knowledge economy but allow research to enhance innovation.

D-1-1. Provide the U.S. Patent and Trademark Office with sufficient resources to make intellectual-property protection more timely, predictable, and effective.	Since 2005, various appropriations acts have provided the United States Patent and Trademark Office with budget authority to spend all fees collected, effectively providing additional resources to this agency.
D-1-2. Reconfigure the U.S. patent system by switching to a "first-inventor-to-file" system and by instituting administrative review after a patent is granted.	Legislation has been introduced to implement "first-to-file," but has not been passed and signed into law.
D-1-3. Shield research uses of patented inventions from infringement liability.	Not acted on.
D-1-4. Change intellectual-property laws that act as barriers to innovation in specific industries, such as those related to data exclusivity (in pharmaceuticals) and those that increase the volume and unpredictability of litigation (especially in information-technology industries).	Various bills, including one passed by the House and one passed by the Senate, would address data exclusivity.
Action D-2: Enact a stronger research and development tax credit to encourage private investment.	Several measures have been passed to bolster the incentive in recent years, and others are under consideration.

TABLE 2-1 Continued

RECOMMENDATIONS AND ACTION STEPS	CONGRESSIONAL ACTIONS
Action D-3: Provide tax incentives for U.S.-based innovation. The Council of Economic Advisers and the Congressional Budget Office should conduct a comprehensive analysis to examine how the United States compares with other nations as a location for innovation.	Not acted on.
Action D-4: Ensure ubiquitous broadband Internet access through spectrum management and regulation.	In 2009, Congress appropriated $7.2 billion (ARRA) for broadband improvement programs. Additional programs and legislation are under consideration.

SOURCE: Adapted from Congressional Research Service, Selected Congressional Actions Related to Recommendations in the 2007 National Academies' Rising Above the Gathering Storm Report, Memo to Senator Jeff Bingaman, February 26, 2010.

NOTE: Appendixes E and F of the original *Gathering Storm* report provide cost estimates for implementing the recommendations. Making a specific assessment of Congressional actions and executive branch implementation against the recommendations would require additional information and analysis. This adaptation is not intended to be comprehensive.

3.0
Changing Circumstances

OVERVIEW OF THE PAST FIVE YEARS

During the five years since the *Gathering Storm* study was published a new research university was established with a "day-one" endowment of $10 billion, equal to what it took MIT 142 years to accumulate.[1] Next year over 200,000 students will study abroad, a large fraction in the fields of science, engineering, and technology.[2] A new "innovation city" is being constructed, patterned after Silicon Valley, that will house 40,000 people.[3] A multi-year initiative is underway to make the country a global nanotechnology hub including constructing 14 new "world-class" universities.[4] A new facility was opened to collect, store, and analyze biological samples and serve as an international hub for biomedical research.[5] A high-level commission with the objective of creating jobs at home has developed a long-term strategy for science and technology patterned after the National Academies study.[6]

---

[1] L. Gold, Skorton and Rhodes attend groundbreaking for Saudi Arabian university, a potential Cornell partner, *Cornell Chronicle*, October 25, *2007 (http://www.news.cornell.edu/stories/Oct07/saudiarabia.html)*; and E. Prentice, MIT Endowment Has 3.2 Percent Yield, Even As U.S. Markets Slide, *The Tech*, October 7, 2008 (http://tech.mit.edu/V128/N45/endowment.html).

[2] W. Wei, Colleges fight for Chinese students, *China Daily*, October 13, 2009.

[3] Smart Russia, *Newsweek*, May 14, 2010. Available at: *http://www.newsweek.com/2010/05/14/smart-russia.html*.

[4] Nanotechnology—The New Frontier for Indian IT Power, Ministry of Foreign Affairs of Denmark, The Trade Council, India, August 8, 2010. Available at: *http://www.eksporttilindien.um.dk/en/servicemenu/News/Nanotechnology TheNewFrontierForIndianITPower.htm*; K. Krishnadas, India Preps Nanotechnology Policy, *EE Times*, December 28, 2007. Available at: *http://www.eetimes.com/electronics-news/4075655/India-preps-nanotechnology-policy*.

[5] For information on the Integrated BioBank of Luxembourg, see: *http://www.ibbl.lu/home/*.

[6] The Royal Society, *The Scientific Century: Securing Our Future Prosperity*, March 2010.

These actions were taken by Saudi Arabia, China, Russia, India, Luxembourg, and the United Kingdom, respectively.

Meanwhile, in the United States, six million more youths dropped out of high school to join a cadre of similarly situated youths—over half of whom under 25 years of age are currently without jobs.[7] During the abovementioned interval, another $2 trillion was spent on K-12 public education while K-12 students remained mired near the bottom of the developed-world class.[8] Labor costs in the United States continue to eclipse those in developing nations, although in some cases by narrowing margins. Over 8.4 million jobs were lost in America . . . and the dollar dropped 9 percent against the Euro.[9] The United States share of global *high-tech* exports dropped from 21 percent to 14 percent while China's share grew from 7 percent to 20 percent.[10] The stock market declined 57 percent before beginning a halting recovery, and some seven trillion dollars of wealth disappeared.[11] The national debt grew from $8 trillion to $13 trillion. Federal debt per citizen increased from $26,700 to $42,000.[12] The nation continued to be dependent upon others for 57 percent of the oil energy it uses.[13] China continued to graduate more *English-trained* engineers than the United States. A number of other countries, including Canada, Australia, and the United Kingdom, conducted their own competitiveness assessments, some patterned after the United States *Gathering Storm* study, and many moved aggressively to implement the actions that were thus identified.[14]

---

[7] Alliance for Excellent Education, High School Dropouts in America, February 2009; Available at: *http://www.all4ed.org/files/GraduationRates_FactSheet.pdf*; A. Sum, I. Khatiwada, J. McLaughlin, and S. Palma, The Consequences of Dropping Out of High School, Northeastern University, October 2009; Available at: *http://www.clms.neu.edu/publication/documents/The_Consequences_of_Dropping_Out_of_High_School.pdf.*

[8] U.S. Department of Education, National Center for Education Statistics, *Statistics of State School Systems, 1969-70; Revenues and Expenditures for Public Elementary and Secondary Education,* 1979-80 and 1980-81; and Common Core of Data (CCD), "National Public Education Financial Survey," 1989-90 through 2006-07, May 2009. Available at: *http://nces.ed.gov/programs/digest/d09/tables/xls/tabn177.xls.*

[9] U.S. Bureau of Labor Statistics, Employees on nonfarm payrolls by industry sector and selected industry detail, August 2010. Available at: *http://www.bls.gov/news.release/empsit.t17.htm* and *ftp://ftp.bls.gov/pub/suppl/empsit.compaes.txt*; Euro-dollar rate was 1.1714 in October 2005, and 1.2703 in August 2010, calculated at *http://finance.yahoo.com.*

[10] National Science Board (NSB), *Science and Engineering Indicators 2010.* Arlington, VA: National Science Foundation (NSB 10-01), Table 6-19.

[11] S&P 500 1565.15 all-time high in October 2007, reached a low of 676.53 in March 2009. $7.1 trillion in wealth based on peak to trough fall in the Wilshire Total Market index, see: *http://www.gurufocus.com/stock-market-valuations.php.*

[12] For national debt, see Table 7.1, Federal Debt at the End of the Year: 1940:2015 at: *http://www.whitehouse.gov/omb/budget/Historicals/* (accessed August 23, 2010); for U.S. population, see *http://factfinder.census.gov/.*

[13] U.S. Energy Information Administration, Petroleum Statistics 2008, Available at: *http://www.eia.doe.gov/basics/quickoil.html.*

[14] The Royal Society, 2010; Panel to Review the National Innovation System, *Venturous Australia,* 2008, see: *http://www.innovation.gov.au/innovationreview/Documents/NIS_review_Web3.pdf*; Science, Technology

Even given these and other recent events, the United States remains relatively strong in comparative economic terms, based in large part on investments made in decades past—such as the GI Bill, post-Sputnik actions to strengthen science and technology, and investments in the Apollo lunar program—the latter of which motivated many young people to pursue careers in science and engineering. U.S. research in the life sciences has produced many life-extending and life-improving medical technologies.[15] Today, with less than five percent of the world's population, the United States still produces 25 percent of global output.[16] It devotes a fraction of GDP to research and development that ranks it among the higher investor nations (although well behind Japan and South Korea).[17] It has a predominant share of the world's finest universities.[18] It maintains the world's strongest conventional and strategic military force. It is the world's largest source of financial capital. It is home to a disproportionate share of the globe's innovators, particularly in high-tech, many of whom immigrated from abroad. Its economic system generally permits weak companies to fail and strong companies to prosper. Other strengths include a unique ability to assimilate immigrants, a research funding allocation system that is largely merit-based (e.g. peer-reviewed competition for nearly all grants), respect for a spirit of adventure and non-conformity, and a willingness to give the best talent independence at an early age. And, stated rather sardonically by Daniel Gross writing in *Newsweek*, "America still leads the world at processing failure."[19]

All of the above is backed by a government characterized by a remarkable degree of stability and operating under the rule of law.

But the continued existence of such assets is not guaranteed—and many, as is becoming increasingly apparent, are subject to atrophy. Indeed, in recent years most competitiveness measures have trended in a flat or negative direction insofar as the United States ability to compete for jobs is concerned.

In addition, three new factors have evidenced themselves during the half-decade that

---

and Innovation Council, *State of the Nation: Canada's Science, Technology and Innovation System*, 2008, see: *http://www.stic-csti.ca/eic/site/stic-csti.nsf/eng/h_00011.html*.

[15] Joint Economic Committee. The Benefits of Medical Research and the Role of the NIH, May 2000. Available at: *http://www.faseb.org/portals/0/pdfs/opa/2008/nih_research_benefits.pdf*.

[16] U.S. Census Population Clock (http://www.census.gov/main/www/popclock.html); NSB, 2010, Appendix Table 6-2.

[17] NSB, 2010, Table 4-11.

[18] Top 200 World Universities, *Times Higher Education*. Available at: *http://www.timeshighereducation.co.uk/Rankings2009-Top200.html*; Academic Ranking of World Universities, *http://www.arwu.org/ARWU2009.jsp*.

[19] D. Gross, The Comeback Country, *Newsweek*, April 9, 2010.

has elapsed since the *Gathering Storm* report was prepared that are particularly significant. Each of these is discussed below.

(1) Decreased Financial Wherewithal to Address the Competitiveness Challenge. While the *Gathering Storm* report warned of an impending financial crisis, it was not addressing the type of crisis that subsequently occurred. It appears that the latter was relatively unique—triggered by government policy that encouraged excessive mortgage borrowing; poor judgment in assessing risk on the parts of both borrowers and lenders; overly aggressive practices by investment banks when creating new financial instruments; and a lack of diligence on the part of regulators. This produced what has been a severe downturn. But it is not the long-term crisis of which the *Gathering Storm* committee sought to warn and avert.

The *Gathering Storm* report sought to call attention to the likelihood of a far more serious and much more enduring financial reversal attributable to fundamental flaws in the nation's process of generating quality jobs for which its citizens can be competitive. This failure includes such practices as tolerating a K-12 educational system that functions poorly in many areas, prolonged underinvestment in basic research, and discouraging talented individuals from other parts of the world, particularly, in science and technology, from remaining in America after having successfully completed their education here.

Although of an altogether different causal character, the recent financial collapse nonetheless impacts the far more profound economic downturn addressed in the *Gathering Storm* report. For example, the funds needed over the long-term to implement and sustain the *Gathering Storm* recommendations will now be substantially more difficult to obtain and sustain with the deficit running at nine percent of GDP and a national debt officially projected to reach $16 trillion by 2020.[20] The current budget plan, if unchanged, is estimated to result in a 70 percent debt-to-GDP ratio in ten years. Further, the trend towards *chronic* unemployment has been accelerated by the downturn . . . with approximately 10 percent of the nation's workforce currently categorized as "unemployed." Infrequently noted, this figure does not include another seven percent that has simply dropped out of the workforce or is "underemployed," which includes those marginally attached to the labor force and those employed part-time for economic reasons.[21] In fact, as the economy eventually recovers from the recent financial market's freeze-up, many of the latter seven

---

[20] Congressional Budget Office, *Budget and Economic Outlook: An Update,* August 2010. Available at: *http://www.cbo.gov/ftpdocs/117xx/doc11705/2010_08_19_SummaryforWeb.pdf.*

[21] U.S. Bureau of Labor Statistics, Alternative measures of labor underutilization, August 2010. Available at: *http://www.bls.gov/news.release/empsit.t15.htm.*

percent are likely to re-enter the job market—thus making the current *true* unemployment rate more nearly equal to 17 percent than 10 percent—further prolonging recovery from the current episode. In June, 2010 there were 32 unemployed individuals competing for each job opening in the construction industry and 7 in manufacturing.[22] Over half the nation's workforce has had its work hours reduced, had its pay cut, been forced to take unpaid leave, or been forced to work part-time during some period since the economic reversal began. A total of 8 million jobs were lost in the downturn, and although the economy has made a substantial turnaround in financial terms, only a small fraction of the lost jobs have been recovered.[23]

In recent decades most new jobs have been generated by smaller firms, many of which only a short time earlier would have been classified as entrepreneurial start-ups. Of the new jobs created in the fifteen-year period prior to the recent economic collapse, 64 percent were produced by small- and medium-sized firms.[24] As those firms revive, unburdened by large prior fixed investments, they will possess much more flexibility than large, established firms to locate new facilities wherever a competitive advantage is to be found.

During the years since the *Gathering Storm* report was produced there has been another change in the character of job creation in America that presumably cannot sustain itself over the longer term. In particular, during this period the private sector eliminated 4,755,000 jobs while government (at all levels) added 676,000 jobs.[25] The difficulty of reversing this trend is exacerbated by yet another development wherein, according to the Bureau of Labor Statistics, federal jobs now pay wages and benefits that on average exceed those in the private sector by 55 percent for similar occupations.[26]

(2) Progress . . . Abroad. While all nations have suffered from the recent financial meltdown, not all have suffered equally. China's GDP grew at an average annual rate of 11 percent between 2005 and 2008; India's by 8.6 percent; Brazil's by 4.5 percent. In contrast, the United States growth rate has averaged 2 percent, albeit from a much larger base but with a much higher standard of living to support.[27]

---

[22] Bureau of Labor Statistics unemployment data available at: *http://www.bls.gov/news.release/archives/empsit_07022010.pdf*; and job openings data at: *http://www.bls.gov/news.release/pdf/jolts.pdf*.

[23] S. Condon, Biden: We Can't Recover All the Jobs Lost, *CBS News*, July 25, 2010, available at: *http://www.cbsnews.com/8301-503544_162-20008924-503544.html*.

[24] Small Business Administration, Office of Advocacy Frequently Asked Questions, available at: *http://www.sba.gov/advo/stats/sbfaq.pdf*.

[25] Calculated from October 2005 to July 2010, at: *http://www.bls.gov/webapps/legacy/cesbtab1.htm*.

[26] D. Cauchon, *Federal Pay Ahead of Private Industry*, *USA Today*, March 8, 2010.

[27] World Bank data available at: *http://data.worldbank.org/indicator/NY.GDP.MKTP.KD.ZG*.

The above circumstance permitted China to increase its R&D investment as a fraction of GDP at an annual rate of 5.7 percent between 2001 and 2007, while the United States investment declined at an annual rate of 0.5 percent.[28] Similarly, the number of first university degrees received in the natural sciences and engineering in China increased at a rate of 42 percent per year whereas the production of such degrees in the United States has increased just 3 percent per year—with part of the increase attributable to growth in the number of non-citizen students receiving degrees.[29]

During the most recent decade China increased its number of higher education institutions from 1,022 to 2,263.[30] During this period the number of students enrolled in degree courses in China increased from about one million to over five million. Tsinghua University, Peking University and Shanghai Jiao Tong University in China and the Indian Institutes of Technology are now considered to be among the world's foremost academic institutions.[31] Perhaps the most innovative of the newly created institutions is KAUST, in Saudi Arabia. KAUST has no departments, no tenure, no undergraduates, no tuition, and a broadly international faculty and student body, heavily focused on research . . . and a very large endowment. It is led by an individual born in Singapore and educated in the United States.[32]

The Information Technology and Innovation Foundation recently analyzed 16 innovation competitiveness indicators and found that the United States now ranks 40th out of the 40 countries and regions considered in "making progress on innovation and competitiveness."[33]

(3) The United States Higher Education Outlook. America is still blessed with a disproportionate share of the world's finest universities—particularly research universities. A recent ranking by *The Times* of London shows six United States universities among the world's top ten. Another ranking by Shanghai Jiao Tong University in China assigns the United States eight of the top ten positions.[34] It is noteworthy that the ranking of institutions of higher education has historically been very resistant to change.

---

[28] NSB, 2010, Appendix Table 4-27.

[29] NSB, 2010, Appendix Table 2-36.

[30] R. Levin, Top of the Class, *Foreign Affairs*, May/June 2010.

[31] Top 200 World Universities, *Times Higher Education*. Available at: *http://www.timeshighereducation.co.uk/Rankings2009-Top200.html*; Academic Ranking of World Universities, *http://www.arwu.org/ARWU2009.jsp*

[32] See: *http://www.kaust.edu.sa/about/admin/president/presidentoffice.html*.

[33] Information Technology and Innovation Foundation, *The Atlantic Century: Benchmarking EU & U.S. Innovation and Competitiveness*, February 2009. See: *http://www.itif.org/files/2009-atlantic-century.pdf*.

[34] Top 200 World Universities, *Times Higher Education*. Available at: *http://www.timeshighereducation.co.uk/Rankings2009-Top200.html*; Academic Ranking of World Universities, *http://www.arwu.org/ARWU2009.jsp*.

Today, however, two forces are at work that could modify that circumstance. The first of these is that a number of other nations are placing extraordinary priority on higher education, particularly in science and engineering. The second is that as a result of the recent financial reversal many United States universities are in greater jeopardy than at any time in nearly a century. As tax revenues have declined, state support of public higher education has been curtailed—in some cases severely. While some argue that the reductions have had the effect of forcing much-needed efficiency measures, in many cases the reductions have gone well beyond improvements in efficiency. In California, home of a world-class set of universities, the state allocation to higher education has, thus far, declined by about 14 percent, triggering what some consider to be Draconian measures of economy—along with a 32 percent increase in tuition.[35] Simultaneously, the endowments of public and private institutions in the United States declined during the recession, suffering an average loss of 18.7 percent during 2008 and 2009.[36]

Companies tend to locate R&D centers near research universities because of the talent and knowledge pools that are locally available. Reductions in America's federal funding for research, coupled with declining state support and shrinking endowments along with the increased stature of foreign universities, can be expected to make U.S. universities less attractive as partners to both established and start-up firms.

The trend towards lesser government funding for public universities in most fields is not new . . . only the magnitude of the decline is new. Exacerbating the problem is the long-standing practice whereby universities receiving federal research grants generally must subsidize these awards. Today, universities are in a precarious position to sustain this practice—particularly those institutions with highly capable but expensive research facilities that must be depreciated. In the words of Paul Courant, James Duderstadt and Edie Goldenberg writing in the *Chronicle of Higher Education*: "[current trends threaten to] cripple many leading public universities and erode their world-class quality."[37] The innovation that is so critical to our economic vitality is in jeopardy when our universities are in jeopardy. In 1975 private firms accounted for more than 70 percent of the "R&D 100" (*R&D* magazine's annual list of the 100 most significant, newly introduced research

---

[35] Legislative Analyst's Office, California. See: *http://www.lao.ca.gov/analysis/2010/highered/Highered_anl10.aspx*; and J. Keller, Amid Protests, U. of California Regents Panel Approves 32% Tuition Increase, *Chronicle of Higher Education*, November 18, 2009. For additional information on tuition increases in other states, see I. Wojciechowska, Paying the Price, Inside Higher Ed, August 27, 2010. Available at: *http://www.insidehighered.com/news/2010/08/27/tuition*.

[36] *NACUBO-Commonfund Study of Endowments, 2009*. Available at: *http://www.nacubo.org/Documents/research/2009_NCSE_Press_Release.pdf*.

[37] P. Courant, J. Duderstadt, and E. Goldenberg, Needed: A National Strategy to Preserve Public Research Universities, *The Chronicle of Higher Education*, January 3, 2010.

and development advances in multiple disciplines), but by 2006, more than 70 percent of the top 100 innovations came from "public or mixed" sources, including academia and federally-supported startups.[38]

Given this demanding environment, a number of other countries are seizing the opportunity to attract United States-educated faculty "superstars" from United States universities where they are now employed. Nobel Laureate Julius Axelrod reminds that "Ninety-nine percent of the discoveries are made by one percent of the scientists."[39] Attracting such individuals to other nations is made easier by political and economic developments in the past two decades that have enabled many more countries to offer reasonable lifestyles along with extraordinary research facilities (e.g., CERN in Switzerland, Biopolis in Singapore, the nuclear-fusion research facilities in China, and the high-energy particle research program in Japan). Further, in the case of engineering, over 35 percent of the faculty of United States institutions was born abroad, considerably easing the disruption of returning home.[40] In this regard the head of R&D for one advanced nation has facetiously referred to himself as a "serial kidnapper."[41]

United States universities, for the first time since World War II, are thus faced with a serious—and increasing—competition for talent from abroad. This has been most noticeable in the case of attracting Japanese students, where undergraduate and graduate enrollments in United States universities have dropped since 2000 by 52 and 27 percent, respectively.[42] The reasons for this decline are manifold—but the result is a measure of the challenge faced as the United States seeks to attract, and keep, talent from abroad. Thus far this particularly severe trend is concentrated among Japanese students, with enrollment in United States universities by Indian, Chinese and South Korean students still increasing.

The percentage of international students who receive doctorates from U.S. universities and are still in the United States two years later declined somewhat during the middle of

---

[38] F. Block and M.R. Keller, *Where Do Innovations Come From? Transformations in the U.S. National Innovation System, 1970-2006,* ITIF, July 2008, available at: *http://www.itif.org/files/Where_do_innovations_come_from.pdf.*

[39] National Academy of Sciences, National Academy of Engineering, Institute of Medicine, *Rising Above the Gathering Storm: Energizing and Employing America for a Brighter Economic Future,* Washington, DC: National Academies Press, 2007, p. ix.

[40] Commission on Professionals in Science and Technology, *The Foreign Born in Science and Technology,* STEM Workforce Data Project. Available at: *https://www.cpst.org/STEM/STEM4_Report.pdf?CFID=11013613&CFTOKEN=27637341.*

[41] B. Walsh, Stem Cell Central, *Time,* July 23, 2006.

[42] B. Harden, Once Drawn to U.S. Universities, More Japanese Students Staying Home, *Washington Post,* April 11, 2010.

the last decade, but the two-year stay rate moved back up to near 70 percent by 2007.[43] The evidence of foreign-born members of the United States academic and industrial workforces returning to their native countries is still largely anecdotal, but focuses on particularly accomplished later-career individuals.

Perhaps the most disconcerting assessment comes from a United States Conference of State Legislatures report: *Transforming Higher Education*, which concludes that "The American higher education system (overall) is no longer the best in the world. Other countries outrank and outperform us."[44]

GLOBAL CHALLENGES IN A GLOBAL WORLD

Other nations are of course not immune to many of the adverse forces addressed herein; indeed, most already had additional serious challenges of their own before the recent economic downturn. Yet, many such nations do not face the task of sustaining the lifestyle which has come to be enjoyed—and expected—by America's citizenry.

India is estimated to have 60 million children suffering from inadequate nutrition.[45] The state-run *China Daily* reports that 150 million Chinese are internal migrant workers from rural areas, who face hardships due to being unregistered in the cities.[46] Fully 2.5 billion of the world's 6.8 billion citizens survive on two dollars per day or less.[47] China's per capita real GDP is one-twelfth that of the United States.[48] As other nations begin to enjoy a higher standard of living their costs of doing business can be expected to rise as well. Indeed, labor costs in China for well educated individuals have increased sharply since the *Gathering Storm* report was conducted, but still fall well below those in the United States. However, due to the enormous labor surpluses in China and India it is unlikely that wages among unskilled workers in those countries will increase significantly in the foreseeable future.

---

[43] M. Finn, Stay Rates of Foreign Doctorate Recipients from U.S. Universities, 2007. Available at: *http://orise. orau.gov/files/sep/stay-rates-foreign-doctorate-recipients-2007.pdf*.

[44] *Transforming Higher Education*, Recommendations of the National Conference of State Legislatures Blue Ribbon Commission on Higher Education, October 2006.

[45] M. Gragnolati, M. Shekar, M. Gupta, C. Bredenkamp and Y. Lee, *India's Undernourished Children: A Call for Reform and Action*, The World Bank, August 2005.

[46] X. Dingding, More help for new migrant workers, *China Daily*, February 2, 2010; A. Scheineson, China's Internal Migrants, Council on Foreign Relations, May 14, 2009.

[47] S. Chen and M. Ravallion, The developing world is poorer than we thought, but no less successful in the fight against poverty, World Bank, August 26, 2008.

[48] U.S. Central Intelligence Agency, *The World Factbook*. See: *https://www.cia.gov/library/publications/the-world-factbook/geos/ch.html* and *https://www.cia.gov/library/publications/the-world-factbook/geos/us.html*.

We live in a time of enormous change—"creative destruction"—a time when new innovations drive out old jobs but create new ones. In the Michigan auto industry, employment plummeted from 460,000 in 1970 to 98,000 today . . . all while jobs derived from efforts in Silicon Valley grew.[49] A nation that does not embrace innovation will soon be left behind in the 21st century economy.

---

[49] M. Guarino, Rising Auto Sales Could Rescue Michigan, Big Three, *The Christian Science Monitor*, March 25, 2010.

4.0

The Ingredients of Innovation

INNOVATION

Given the factors previously discussed that militate against creating (or preserving) jobs in the United States, how then is America to maintain, or preferably enhance, the future standard of living of its citizenry? The answer (and seemingly the only answer) is through *innovation*. "Innovation" commonly consists of being first to acquire new knowledge through leading-edge research; being first to apply that knowledge to create sought-after products and services, often through world-class engineering; and being first to introduce those products and services into the marketplace through extraordinary entrepreneurship.

Writing in *Foreign Affairs*, Yale President Richard Levin notes:

"To oversimplify, consider the following puzzle: Japan grew much more rapidly than the U.S. from 1950 to 1990, as its surplus labor was absorbed into industry, and much more slowly than the United States thereafter. Now consider if Japan would have grown so slowly if Microsoft, Netscape, Apple and Google had been Japanese companies. Probably not. It was innovation based on science that propelled the United States past Japan during the two decades prior to the crash of 2008. It was Japan's failure to innovate that caused it to lag behind."[1]

In the words of Wharton Professor Jeremy Siegel, "Economic growth is based on advances in productivity, and productivity is based on discovery and innovation."[2]

---

[1] R. Levin, Top of the Class, *Foreign Affairs*, May/June 2010.
[2] The Shape of Things to Come, *Newsweek*, April 8, 2010. Available at: *http://www.newsweek.com/2010/04/08/the-shape-of-things-to-come.html*.

Speed is of the essence in introducing innovation in a competitive economy. Craig Barrett, the retired chairman and CEO of Intel Corporation, states that 90 percent of the revenues that firm derives on the last day of the year are attributable to products that did not even exist on the first day of that same year.[3] Fortunately, Americans, especially immigrant Americans, have been demonstrated to be very accomplished innovators. They are commonly risk-takers.

The primary ingredients of successful innovation can thus be categorized as (1) new knowledge; (2) capable people, and (3) an environment that promotes innovation and entrepreneurship. Each of these factors is discussed, and the United States position assessed, in the succeeding sub-sections under the respective headings, "Knowledge Capital," "Human Capital," and the "Environment."

4.1 KNOWLEDGE CAPITAL

The most fundamental building block of innovation is newly acquired knowledge, often in the form of scientific or technological advancements. Margaret Thatcher observed that,

> . . . although basic science can have colossal economic rewards, they are totally unpredictable. And therefore the rewards cannot be judged by immediate results. Nevertheless, the value of Faraday's work today must be higher than the capitalization of all shares on the stock exchange. . . . The greatest economic benefits of scientific research have always resulted from advances in fundamental knowledge rather than the search for specific applications . . . transistors were not discovered by the entertainment industry . . . but by people working on wave mechanics and solid state physics. [Nuclear energy] was not discovered by oil companies with large budgets seeking alternative forms of energy, but by scientists like Einstein and Rutherford. . . .[4]

Unfortunately, the very real pressures of today's financial markets make it difficult for corporations to invest in fundamental research, which by its very nature is risky, long-term, of uncertain applicability, and increasingly expensive—the latter particularly in the United States. In one survey, 80 percent of chief financial officers of United States firms respond-

---

[3] N.R. Augustine, *Is America Falling Off the Flat Earth?* Washington, DC, National Academies Press, 2007.
[4] M. Thatcher, Speech to the Royal Society, September 27, 1988. Available at: *http://www.margaretthatcher. org/speeches/displaydocument.asp?docid=107346*; M. Kenward, Let's Get Back to Basics, Says Thatcher, *New Scientist*, December 16, 1989. Available at: *http://www.newscientist.com/article/mg12416950.900-lets-get-back-to-basics-says-thatcher.html*.

ing indicated they would cut R&D to meet their firm's next-quarter's profit projections.[5] Pharmaceutical companies report that only one of every ten thousand chemicals they investigate as potential new medicines is ultimately approved for patient use. According to one estimate, it costs on average $802 million, an amount that continues to increase and takes an average of 12 years, to transition one new chemical from the exploratory phase to use by United States patients.[6] Such considerations represent a great barrier to investors, both large and small.

In this environment the great United States corporate research laboratories of the past are increasingly becoming a thing of the past. The canonical case is Bell Laboratories, home of the transistor, the laser and numerous Nobel Laureates—which was gradually downsized until the remainder was sold to a French firm. As other nations have increased their investments in research, discoveries can be expected to shift abroad as well. For example, the development of new research tools is an important by-product of the research process. Successful innovation requires the invention of new tools that allow for more precise measurements, the production of purer or better materials, and more effective manipulation of data. A case in point is the field of particle physics which employs high energy accelerators as a principal discovery tool. Since their invention, the most capable of these machines has always been located in the United States—until recently when, for the first time, the most capable machine is located abroad, in Switzerland and France.

Given the trend of industry to invest less in fundamental research, focusing on more predictable development projects, it is increasingly left to government to fund the former type of activity. This is consistent with the notion that governments *should* assume responsibility for supporting activities that produce benefits to society as a whole but not necessarily commensurately to the individual performer or underwriter. In such a scenario the nation's research universities will have to assume even greater responsibility for *performing* much of the nation's research—with that research largely being funded by the federal government. In 2008, about 43 percent of the $68 billion worth of research (basic and applied) supported by various federal agencies was performed at universities.[7] It is noteworthy that such activity is rapidly becoming globalized, with the percentage of

---

5 J. Graham, C. Harvey, and S. Rajgopal, The Economic Implications of Corporate Financial Reporting, September 13, 2004. Available at: *http://faculty.fuqua.duke.edu/~charvey/Teaching/BA456_2006/The_economic_implications.pdf.*

6 Tufts Center for the Study of Drug Development, How New Drugs Move through the Development and Approval Process, November 2001.

7 National Science Board (NSB), *Science and Engineering Indicators 2010.* Arlington, VA: National Science Foundation (NSB 10-01), Appendix Tables 4-8 and 4-9.

internationally co-authored research articles almost tripling between 1998 and 2008.[8] A concern going forward is the current increasing investment by the National Institutes of Health in translational (or applied) research focusing more on drug discovery as opposed to the generation of new, fundamental knowledge which is the limiting factor in true innovation.

One common measure of scientific *input* is the fraction of a nation's GDP that is devoted to scientific research . . . on the principle that the size of the economy to be maintained affects the size of the effort needed for its maintenance. Given similar research efficiencies among nations, which of course may or may not be the case, this factor should correlate directly (not necessarily linearly) with research *output*. By this measure, basic research as a fraction of GDP, the United States most recently ranked fifth among all nations.[9]

Turning to research *and* development—where the United States ranks eighth among nations on a per-GDP basis—government investment has declined from two-thirds of the nation's total expenditure to less than one-third.[10] Over half of United States federal R&D spending is defense-related. China has a relatively low R&D to GDP ratio—but has more than doubled the figure over the past decade, even while growing its GDP substantially.

Viewing such trends United States research universities are increasingly creating ties to what they view as the more highly regarded overseas universities. For example, the president of Yale University has cited the benefits being realized from a partnership in the biosciences with a Chinese university. In that case, a competitive edge was derived by sending researchers to China rather than having them come to the United States because of the lower costs, excellent facilities and abundance of talented graduate students in China.

United States industrial firms are increasingly adopting much the same strategy, building new research facilities outside the country. Although this was initially driven by the lower cost of operations abroad, it now is often motivated by the relative availability of talent. The National Science Foundation reports that U.S.-based companies now have 23 percent of their R&D employment located abroad.[11]

---

[8] NSB, 2010.

[9] NSB, 2010, Table 4-12.

[10] NSB, 2010, Table 4-11.

[11] F. Moris and N. Kannankutty, New Employment Statistics from the 2008 Business R&D and Innovation Survey, National Science Foundation, July 2010. Available at: *http://www.nsf.gov/statistics/infbrief/nsf10326/nsf10326.pdf*.

4.2 HUMAN CAPITAL

Workforce Education

The chairman of the Department of Commerce's National Advisory Committee on Measuring Innovation, Carl Schramm, notes that "Nobel Laureate Gary Becker developed a theory that empirically established that people were more important to an economy than physical capital. Becker's now obvious observation is central to conscious attempts to induce more innovation. In his book *The Vital Few,* economist theorist Jonathan Hughes points out that the welfare of society, connected as it is to innovation and entrepreneurship, hangs on a very small number of our fellow citizens."[12]

Dean Yash Gupta of the Johns Hopkins Carey Business School further notes that ". . . 30 years ago the United States had 30 percent of the world's college students. Today we are at 14 percent and falling. Twenty years ago the U.S. was first among industrialized nations in share of population with a high school diploma and first with a college degree. Today, we are ninth in high school diplomas (and) seventh in college degrees worldwide. We are 18th out of 24 in high school graduation (rate) among industrialized nations . . . and falling."[13] At the same time, employers indicate that knowledge demands on all employees are higher than ever. A recent case reported in *The New York Times* stated that a firm seeking to hire employees was able to find only 47 who were qualified out of an applicant pool of 3,600.[14]

Science, Engineering and Mathematics

It has increasingly become recognized that to be competitive in the global technology-dominated marketplace requires a highly qualified workforce. This in turn demands that virtually all job-seekers be at least "proficient" in mathematics and general science and that the nation have a cadre of highly creative individuals who possess an extraordinary capacity for mathematics, science and engineering.

It is not necessary—or even possible—to seek to match nations such as China and

---

[12] Carl Schramm, Made in America, *The National Interest*, April 2010. Available at: *http://www.usinnovation.org/files/SchrammMadeinAmerica.pdf.*

[13] Y. Gupta, Innovation: Can a Nation Have a Second Act? Speech to the Baltimore Rotary, June 8, 2010. Figures in this quote may differ from similar indicators cited in other parts of this report due to different sources or coverage in terms of dates, degrees (four-year vs. combined two- and four-year) or countries.

[14] M. Rich, *Factory Jobs Return, but Employers Find Skills Shortage,* The New York Times, July 1, 2010.

India, each with approximately four times the population of the United States, in over-all quantities of scientists and engineers. Further, the race for quantity has already been rather decisively lost. Jobs performing relatively routine functions of science and engineering have been lost to nations with lower cost structures and a well educated citizenry. What must be preserved in the United States, if the nation is to compete, is an adequate supply of scientists and engineers who can perform creative, imaginative, leading-edge work—that is, who can *innovate*. Albert Einstein wrote, "Imagination is more important than knowledge. Knowledge is limited. Imagination encircles the world."[15]

The principal focus of the *Gathering Storm* review was on mathematics, science and engineering, not simply because of their critical importance in creating jobs but also because these are the disciplines in which American education is failing most convincingly. This is not to diminish the importance of many other fields—particularly reading at the elementary school level and the liberal arts in all grades. Nor does it overlook the fact that there is indeed a limited number of truly extraordinary public schools in America. It merely recognizes that it is difficult to dismiss evidence such as the survey that found that almost 30 percent of American adults do not know the earth revolves around the sun; 16 percent do not know that the center of the earth is very hot; almost half do not know that electrons are smaller than atoms; and only about half the population is aware that dinosaurs and humans never coexisted.[16]

Production of Scientists and Engineers

In spite of the nation's growing population and the explosion of knowledge in science and technology and its impact during the past decade, the number of recipients of bachelor's degrees in mathematics, engineering and the physical sciences from United States universities has remained virtually unchanged.[17]

The numbers of doctorate degrees awarded by United States universities in mathematics and the physical sciences have likewise remained basically unchanged in the past decade.[18] In contrast, the number of engineers receiving doctorates has evidenced significant growth (from about 6,000 to about 8,000 graduates per year) in the last five years for which data are available. The increase is, however, largely attributable to the growth of

---

[15] K. Taylor, Is Imagination More Important Than Knowledge? *Times Higher Education*, October 2, 2002.
[16] NSB, 2010, Appendix Table 7-10; National Science Board (NSB), *Science and Engineering Indicators 2002*, Arlington, VA: National Science Foundation, Chapter 7.
[17] NSB, 2010, Figure 2-5.
[18] NSB, 2010, Figure 2-14.

foreign student enrollment.[19] As a basis of comparison, United States universities award about 150,000 MBA's, 44,000 law degrees, 68,000 engineering (undergraduate) degrees and 8,000 engineering PhD's each year.[20]

While the representation of women among those receiving bachelor's degrees in *all* fields from United States universities exceeds 57 percent, less than 20 percent of the degrees in engineering are awarded to women—with the most recent trend slightly worsening.[21] Among sixth graders who received scores above 700 on the mathematics Scholastic Aptitude Test thirty years ago boys outnumbered girls by 13:1. In the more recent tests, the ratio is 4:1—suggesting once again that societal rather than biological issues are at work here.[22] Similarly, black and Hispanic representation among those receiving bachelor's degrees in engineering is less than one-half their proportionate share of the overall population. The situation in the physical sciences is somewhat more balanced than in engineering, but still unbalanced.

The comparative underrepresentation of United States citizens studying the natural sciences and engineering, particularly at the doctoral level, is of particular concern. Students receiving their undergraduate degrees in the natural sciences or engineering from United States undergraduate institutions represent 16 percent of total enrollment of those institutions. This contrasts with 47 percent in China, 38 percent in South Korea, and 27 percent in France.[23]

Overall, 47 percent of U.S. four-year college students fail to graduate within six years.[24] Over the last decade or so the United States has fallen from first to 16th in tertiary graduation rate.[25]

A paradox exists in the debate over whether there is a shortage of scientists and

---

[19] NSB, 2010, Appendix Table 2-28.

[20] NSB, 2010, Appendix Tables 2-12 and 2-28; American Bar Association. See: *http://www.abanet.org/ legaled/statistics/charts/stats%20-%201.pdf*; and Association to Advance Collegiate Schools of Business, Business School Data Trends and 2010/List of Accredited Schools. Available at: *http://www.aacsb.edu/ publications/businesseducation/2010-Data-Trends.pdf*

[21] NSB, 2010, Appendix Table 2-12.

[22] K.L. Bates, Gender Gap in Math Scores Persists, Duke University News Office, July 2, 2010. Available at: *http://news.duke.edu/2010/07/TIPability.html*.

[23] NSB, 2010, Appendix Table 2-35.

[24] M.B. Marklein, 4-year colleges graduate 53% of students in 6 years, *USA Today*, June 3, 2009. Available at: *http://www.usatoday.com/news/education/2009-06-03-diploma-graduation-rate_N.htm*.

[25] Organization for Economic Cooperation and Development, *Education at a Glance 2009: OECD Indicators*, Paris, 2009. Rankings include OECD members and partners, and college graduation ranking is based on Tertiary-A institutions. See: Chart A3.2 in *http://www.oecd.org/document/24/0,3343,en_2649_39263238_43586328_1_1_1_1,00.html*.

engineers or whether there are too many scientists and engineers for the jobs that are available. Most business leaders maintain the former; however, with regard to the more "conventional" functions of these fields it may well be that *de facto* there can no longer be domestic shortages of scientists and engineers. Firms facing this proposition are simply moving work elsewhere. Similarly, the observation that many scientists and engineers elect to pursue careers in other fields is in many instances simply reflective of the value placed on education in these disciplines by business, law, and medical schools and related employers and should not necessarily be decried. However, if the sole purpose of a PhD in science is considered to be to prepare future educators in science, then a surplus of scientists (often evidenced as a surplus of Post-Doctorate researchers) seems inevitable. The *Gathering Storm* recommendations are based upon the premise that federal investment in research must be doubled (the report's second highest priority recommendation)—in which case there will be commensurate increases in demand for researchers . . . and not solely for the purpose of providing educators. Further, since only about four percent of the U.S. workforce is engaged in science and engineering, even rather large increases in employment in these disciplines will have only a modest direct impact on overall employment. It is the leverage in jobs that these individuals create for others to which their value is attributable.

K-12 Education

About thirty percent of United States youths fail to receive a high school diploma on time.[26] The United States is now 20th in high school graduation rate among industrialized nations.[27] One consequence is that, according to a recent report, 75 percent of United States youth are ineligible for service in the nation's military due to academic, physical or moral shortcomings.[28] In July 2010, the unemployment rate among those of all ages who did not complete high school was 13.8 percent, whereas it was 10.1 percent among high school graduates, 8.3 percent among those with some college, and 4.5 percent among those with at least a bachelor's degree.[29] Significantly, only one in 17 children from fami-

---

[26] U.S. Department of Education, National Center for Education Statistics, *High School Dropout and Completion Rates in the United States: 2007*, September, 2009. Available at: http://nces.ed.gov/pubs2009/2009064. pdf; see also White House Press Release, President Obama Announces Steps to Reduce Dropout Rate and Prepare Students for College and Careers, March 1, 2010. Available at: *http://www.whitehouse. gov/the-press-office/president-obama-announces-steps-reduce-dropout-rate-and-prepare-students-college-an*.

[27] OECD, 2009. Rankings include OECD members and partners. See: Chart A2.1 in *http://www.oecd.org/ document/24/0,3343,en_2649_39263238_43586328_1_1_1_1,00.html*.

[28] C. Davenport and E. Brown, Girding for an Uphill Battle for Recruits, *Washington Post*, November 5, 2009.

[29] Bureau of Labor Statistics data available at*: http://www.bls.gov/news.release/empsit.t04.htm*.

lies with less than $35,000 annual income obtain a bachelor's degree by age 24.[30] It is these children from low-income homes that are most under-served by the nation's public school system.

Even those students who do graduate from high school often find a chasm between the requirements for a high school diploma and what is needed to succeed in college— one result of which is that 77 percent of college freshman are unable to pass a college preparatory examination in at least one of three core subjects, and about one third have to take remedial work in college.[31] Students requiring remediation graduate from college at a much lower rate compared with those who do not—a very costly attempt at a solution to the nation's K-12 shortcomings—although it should be noted that some of this disparity is undoubtedly due to financial considerations.[32]

In international standardized tests involving students from 30 nations, United States fourteen-year-olds rank 25th in mathematics and 21st in science.[33] In tests *within* the United States, little improvement has been observed over the past 40 years. This is in spite of a sevenfold increase in inflation-adjusted spending per student since World War II.[34] More recently, in 1971 per-student K-12 spending was $4,489; in 2007 the corresponding figure, adjusted for inflation, was $10,041.[35] In 1973 the average score on one standardized test (the National Assessment of Education Progress) in mathematics among 17-year-olds was 304 out of 500. A third of a century later it was 306.[36] In reading, the corresponding gain in the scores was from 285 to 286.[37] In the most recent test, three jurisdictions out of 51 (50 states plus the District of Columbia) showed significant improvement in fourth grade reading, while 44 showed essentially no gain and four showed marked declines.[38] Among high school seniors average scores in the National

---

[30] R.D. Kahlenberg, Cost Remains a Key Obstacle to College Access, *Chronicle of Higher Education*, March 10, 2006.

[31] ACT Policy Report, Courses Count: Preparing Students for Postsecondary Success, Available at: *http://www. act.org/research/policymakers/pdf/CoursesCount.pdf.*

[32] National Center for Education Statistics, Remediation and Degree Completion, 2004. Available at: *http:// nces.ed.gov/programs/coe/2004/section3/indicator18.asp.*

[33] NSB, 2010, Appendix Table 1-11.

[34] A. Peng and J. Guthrie, The Phony Funding Crisis, Education Next, Winter 2010. Available at: *http:// educationnext.org/the-phony-funding-crisis/.*

[35] National Center for Education Statistics, *Digest of Education Statistics, 2009* (NCES 2010-013). Available at: *http://nces.ed.gov/fastfacts/display.asp?id=66.*

[36] The Nation's Report Card, Trend in NAEP mathematics average scores for 17-year-old students. Available at: *http://nationsreportcard.gov/ltt_2008/ltt0002.asp?subtab_id=Tab_3&tab_id=tab1#chart.*

[37] Trend in NAEP reading average scores for 17-year-old students. Available at: *http://nationsreportcard. gov/ltt_2008/ltt0003.asp?subtab_id=Tab_3&tab_id=tab1#chart.*

[38] National Center for Education Statistics, *Reading 2009: National Assessment of Educational Progress at Grades 4 and 8*, March 2010. Available at: *http://nces.ed.gov/nationsreportcard/pdf/main2009/2010458.pdf.*

Assessment of Educational Progress have actually declined during the most recent decade for which data are available in science.[39] Indications of very recent improvements in some isolated cases are now being questioned as an artifact of changing examination rigor. As but one example, in New York State eighth graders reaching the "proficiency" standard increased from 59 to 80 percent between 2007 and 2009, while the same group's scores on the national math test remained virtually unchanged.[40] This is a phenomenon which is by no means unique to New York State.

McKinsey & Company, the management consultant, concluded in a recent study that disparities in U.S. K-12 education compared to those of many other nations "impose the economic equivalent of a permanent national recession—one substantially larger than the deep recession the country is currently experiencing."[41]

The average student intending to major in education in United States universities ranks in the 42nd percentile of all students taking the college boards in Critical Reading, in the 41st in mathematics and in the 46th in writing.[42] An international test in mathematics content knowledge at the lower secondary level, involving teachers nearing the end of their college education, ranked United States future teachers in seventh place among the 15 nations that participated.[43]

Forty-six percent of teachers abandon their profession within five years of first entering the classroom.[44] Yet, according to *The New York Times,* when the city of New York invested $2 million in additional lawyers to assist in discharging teachers considered to

---

[39] National Center for Education Statistics, *Science 2005: National Assessment of Educational Progress at Grades 4, 8 and 12,* May 2006. Available at: *http://nces.ed.gov/nationsreportcard/pdf/main2005/2006466. pdf.*

[40] New York State Department of Education, A New Standard for Proficiency: College Readiness (Slide Presentation), July 28, 2010. Available at: *http://www.oms.nysed.gov/press/PressConferencePresentation UPDATEDAM07_28.pdf.* Note that for 2010 the score needed to achieve Level 3 proficiency in math was raised for eighth graders, so while the results were flat, the proportion achieving proficiency declined to 55 percent.

[41] M. Hirsh, We're No. 11! America May Be Declining, But Don't Despair, *Newsweek,* August 23/30, 2010.

[42] The College Board, *2008 College-Bound Seniors: Total Group Profile Report.* Available at: *http:// professionals.collegeboard.com/profdownload/Total_Group_Report.pdf;* The College Board, *SAT Percentile Ranks, 2009 College-Bound Seniors.* Available at: *http://professionals.collegeboard.com/profdownload/SAT-Percentile-Ranks-2009.pdf.*

[43] The Center for Research in Math and Science Education, Michigan State University, *Breaking the Cycle: An International Comparison of U.S. Mathematics Teacher Preparation,* Initial Findings from the Teacher Education and Development Study in Mathematics, 2010. Available at: *http://www.educ.msu.edu/content/sites/usteds/ documents/Breaking-the-Cycle.pdf.*

[44] G. Saitz, *Growing Great Teachers,* National Education Association website. Available at: *http://www.nea. org/home/37001.htm.*

be incompetent, the effort removed three teachers (out of a total of 55,000) over the past two years.[45]

McKinsey & Company concludes that if United States youth could match the performance of students in Finland, America's economy would be between nine and sixteen percent larger.[46] That equates to between 1.3 and 2.3 *trillion* dollars each year.

It should be reiterated that the need to strengthen science and math education in the nation's public schools is not simply to produce more graduates possessing the qualifications needed to pursue degrees and careers in science and engineering. The spectrum of jobs that is available to high school as well as college graduates is increasingly demanding at least rudimentary skills in these fields.

Importing Talent

A logical question is how United States science and engineering has managed to prosper with such a tenuous underpinning. A substantial part of the answer is that the United States has benefited immensely from, and is highly dependent upon, foreign-born individuals talented in science and engineering who elect to study in the United States and decide to remain here after completing their education. It probably would not be an overstatement to assert that America's science and engineering enterprise would barely function without these talented contributors.

Of the PhDs in the United States science and engineering workforce under the age of 45—considered to be the most productive years in science—35 percent are foreign-born.[47] Thirty-five percent of United States engineering faculty is foreign-born and 57 percent of "post-docs" in this country are temporary residents.[48] Forty-six percent of the members of the United States physics team and 65 percent of the top United States scorers in the Mathematics Olympiad are the children of immigrants.[49]

---

[45] J. Medina, Progress Slow in City Goal to Fire Bad Teachers, *The New York Times*, February 23, 2010.

[46] McKinsey & Company. *The Economic Impact of the Achievement Gap in America's Schools.* April 2009. Available at: *http://www.mckinsey.com/App_Media/Images/Page_Images/Offices/SocialSector/PDF/achievement_gap_report.pdf.*

[47] National Science Foundation, *Characteristics of Doctoral Scientists and Engineers in the United States: 2006,* September 2009. Available at: *http://www.nsf.gov/statistics/nsf09317/pdf/nsf09317.pdf.*

[48] National Science Foundation, *Graduate Students and Postdoctorates in Science and Engineering: Fall 2007,* June 2010. Available at: *http://www.nsf.gov/statistics/nsf10307/pdf/nsf10307.pdf.*

[49] S. Anderson, The Multiplier Effect, *International Education,* Summer 2004. Available at: http://www.nfap.net/researchactivities/studies/TheMultiplierEffectNFAP.pdf.

The contribution of foreign-born individuals is not limited to basic research: Yahoo, Sun Microsystems, eBay, Intel and Google were all founded or co-founded by immigrants from Taiwan, Germany, India, France, Hungary, or Russia. During the 10 years following 1995, 52 percent of Silicon Valley start-ups—now employing many thousands of people—were founded by immigrants.[50] According to a Duke University study, foreign-born entrepreneurs during the period from 1995 to 2005 founded, or were partners in founding, one-fourth of the new engineering and technology companies in the United States, employing 450,000 workers in 2005. A strong multiplier effect exists when creative scientists and engineers are provided an innovation-friendly environment. Yet, United States immigration policy in many cases discourages qualified individuals from studying in the United States or remaining here after graduation.

As the rest of the world enjoys increasing prosperity and greater freedom some foreign-born graduates of United States universities are being attracted to return home. Although this trend is not massive at this point, there are numerous specific examples relating to some of America's more renowned researchers. A recent change of attitudes is indicated in a Kauffman Foundation survey that found a majority of Indian and Chinese students indicating they would like to remain in the United States a "few" years after graduation, but only six and ten percent, respectively, said they would like to remain permanently.[51] Once a tipping point has been reached in a nation's ability to innovate, the decline becomes self-reinforcing as students no longer seek to attend that nation's universities and graduates seek work in more promising venues.

4.3 ENVIRONMENT

The Innovation Ecosystem

Once new research discoveries have been converted into products and services through the application of advanced engineering practices it becomes the role of entrepreneurs to assure that those products and services are first to market. Even weeks can matter in the race to be first; hence, the job-creating value of research is highly perishable. It took

---

[50] V. Wadhwa, Foreign-Born Entrepreneurs: An Underestimated American Resource, *Kauffman Thoughtbook 2009*, The Kauffman Foundation. Available at: *http://www.kauffman.org/entrepreneurship/foreign-born-entrepreneurs.aspx*.

[51] V. Wadhwa, A. Saxenian, R. Freeman, and A. Salkever, *Losing the World's Best and Brightest: America's New Immigrant Entrepreneurs*, Part V, March 2009. Available at: *http://www.kauffman.org/uploadedFiles/ResearchAndPolicy/Losing_the_World's_Best_and_Brightest.pdf*.

almost two years for one million iPods to be sold; 74 days for one million iPhones; and 28 days for one million iPads.[52] Any ecosystem that delays, or worse yet, halts, the passage of ideas into products and then into markets can undermine the entire innovation process.

One impediment to being first to market is often referred to as "the Valley of Death"— actually not one but several valleys. Prominent among these is the situation where a product, not yet free of significant risk but offering considerable promise, demands substantial additional investment to take the next (costly) step towards the marketplace. The operative question is what might be a source of funds under such a circumstance and in today's economic environment risk-capital is very difficult to obtain.

Collaboration across sectors is absolutely critical to sustain innovation in some industries. For example, one study found that 31 percent of new products and 11 percent of new processes in biomedical fields could not have been developed or would have been significantly delayed without contributions from academic research.[53]

The "innovation ecosystem" thus refers to that set of circumstances which assist—or inhibit—the innovation process. Some of the more critical elements of this ecosystem include:

Cost of Labor

As noted previously, labor costs continue to be considerably higher in the United States than in the less developed parts of the world. For example, nearly twenty assembly workers can be employed in Vietnam for the cost of one in the United States. As other nations prosper these differences will presumably diminish; however, it can be expected to take considerable time before anything approaching parity is reached. In the case of China, the trend towards greater wealth in the cities and in certain suburbs is having an equalizing impact and some economists predict China's advantage in the cost of manufacturing labor will have considerably diminished within two decades. Nonetheless, the United States is likely to endure a not insignificant *overall* labor cost disadvantage for many years.

---

[52] Apple Sells One Millionth iPhone, Press Release, September 10, 2007; Apple Sells One Million iPads, Press Release, May 3, 2010.

[53] E. Mansfield, Academic Research and Industrial Innovation, *Research Policy* 20 (1991): 1-12.

Tort Policy

As previously noted, United States firms spend over twice as much on litigation as on research—a realm unapproached anywhere else in the world.[54] This tends to discourage even prudent risk-taking and consumes resources and vast amounts of time and imposes a severe opportunity-cost. Similarly, the judicial process for resolving disputes tends to be cumbersome and time consuming, such that firms often have no choice but to settle even frivolous cases if they are to avoid still further damage. This legal process often consumes years or even decades to arrive at a resolution to a dispute, yet many businesses are born, prosper and sometimes fail in five years or less.

Tax Policy

The United States has the second highest corporate tax rate among industrialized nations, exceeded only by Japan, backed by 17,000 pages of regulations and interpretations.[55] Although the United States once offered the most generous R&D tax credit in the world, it now ranks 17th of 30 OECD countries.[56] It is not uncommon for United States firms to be attracted abroad by highly preferential tax rates that are offered to relocating firms. Given the growing national debt being assumed by the United States, tax rates can be expected to become an increasingly significant factor in considering where to start or expand a business.

Regulatory Barriers

Well-intentioned regulations can and often do have important unintended consequences. One growing trend among start-ups is to initiate business in Canada because of the greater ease of founding, licensing, operating and selling a firm.[57] In the case of

---

[54] NSB, 2010, Appendix Tables 4-8 and 4-9; Towers Perrin, *2009 Update on U.S. Tort Cost Trends*, Appendixes 1-5.

[55] S. Hodge, U.S. States Lead the World in High Corporate Taxes, Tax Foundation Fiscal Fact 119, March 18, 2008. Available at: *http://www.taxfoundation.org/publications/show/22917.html*; I. Greenwald, High Corporate Tax Rate Is Misleading, *Smart Money*, January 25, 2008. Available at: *http://www.smartmoney.com/investing/economy/high-corporate-tax-rate-is-misleading-22463/*; Editorial: Staggering Facts About Our Tax Code, *The Lima News*, April 15, 2009. Available at: *http://www.www.istockanalyst.com/article/viewiStockNews/articleid/3194996.*

[56] R. Atkinson and S. Andes, *U.S. Continues to Tread Water in Global R&D Tax Incentives*, Information Technology and Innovation Foundation, August 13, 2009.

[57] The World Bank Group ranks Canada as the second best country in the world to start a business, after New Zealand. Rankings available at: *http://www.doingbusiness.org/economyrankings/.*

the Food and Drug Administration, which regulates 25 cents of every dollar spent by the average American, or over a $1 trillion in products ranging from cosmetics to pet foods to medical products, resources have not kept pace with responsibilities.[58] The lack of certainty and predictability in the review and approval process heightens risks of failure, raises the costs of development, and makes the struggle for capital still more difficult.

Cost/Availability of Capital

The United States has long enjoyed a major advantage in terms of the availability of venture capital: California has had more venture capital available than any of the world's *nations* (excluding, of course, the United States, making up about half of the United States total).[59] Today, investors, including an increasing number abroad, are placing less and less emphasis on geopolitical borders as they search for the opportunity for financial gain. As a result, the "source of capital" advantage enjoyed by the United States in the past has been diminishing. The nation's unique advantage is being eroded by the rise of venture financing elsewhere.[60] This problem has been particularly acute for smaller firms where the availability of risk capital has greatly contracted.

Protection of Intellectual Capital

Much of the world's business depends upon the United States patent system for the protection of intellectual property. Nonetheless, that system is ponderous and glacial, in part due to a shortage of an adequate number of qualified personnel and the system's heavy dependence upon litigation. Only 26 percent of the patent examiners reviewing "business methods" patents have any industry experience.[61] Less than half the judges in the "specialized" patent court have technical backgrounds.

---

[58] See Subcommittee on Science and Technology, FDA Science and Mission At Risk, November 2007. Available at: *http://www.fda.gov/ohrms/dockets/ac/07/briefing/2007-4329b_02_00_index.html.*

[59] National Venture Capital Association data available at: *http://www.nvca.org/index.php?option= com_content&view=article&id=78&Itemid=102.*

[60] The United States still accounts for about two-thirds of global venture investments, but venture investing in China, India, and elsewhere in Asia is growing rapidly. See Dow Jones Venture Source, Global Venture Investment Rises 13% in First Quarter of 2010-Press Release, April 29, 2010.

[61] K. Teska, Who Makes the Patent Calls? *Mechanical Engineering*, April 2010. Available at: *http://memagazine. asme.org/Articles/2010/April/Makes_Patent_Calls.cfm.*

Freedom from Corruption

One of the more powerful factors in deterring responsible firms from building plants, creating jobs and conducting business in any nation is the prevalence of corruption. Russia is often cited as an example of such a nation and its economy has suffered accordingly. It is noteworthy that in 2009 the United States ranked 19th in the world on Transparency International's Corruption Perception Index, where the higher the rank the less corruption is perceived.[62] The index includes such practices as bribery, price-fixing, employee theft, and misrepresentation of financial condition.

Sanctity of Law

In general the United States system for enforcing contracts, resolving disputes, assuring human safety, protecting property, caring for the environment and dealing with related issues is widely considered to be superior to that of many nations, particularly developing nations. However, as previously noted it is a costly and deliberate system to the point of incompatibility with today's fast-moving global commerce.

Cost of Benefits

United States standards for the provision of benefits to employees by corporations (pensions, healthcare, savings accounts, vacations, holidays, etc.) are substantially more generous—i.e., more costly—than is the case in the less-developed nations (but significantly less than in Europe). Such benefits, often costing employers one-third of wages or more, perform an important societal function; nonetheless, their existence must be reflected in the price of a firm's products. This, in turn, impacts competitiveness. In the United States, General Motors spends more on healthcare than on steel; Starbucks more on healthcare than on coffee.[63]

---

[62] Transparency International, Corruptions Perception Index 2009. Available at: *http://www.transparency. org/policy_research/surveys_indices/cpi/2009/cpi_2009_table.*

[63] J. Appleby and S. Carty, Ailing GM looks to scale back generous health benefits, *USA Today*, June 24, 2005. Available at: *http://www.usatoday.com/money/autos/2005-06-22-gm-healthcare-usat_x.htm;* Health care takes its toll on Starbucks, *Associated Press*, September 14, 2005. Available at: *http://www.msnbc.msn. com/id/9344634/.*

Export Control Laws

The ability to move products and knowledge into and out of the United States is controlled by the ITAR (International Traffic in Arms Regulation), Export Controls, and Deemed Export Controls. Most of these laws and regulations were promulgated at a time when the United States held a dominant position in technology and when transferring militarily-significant knowledge out of the country was extremely difficult—neither of which pertains today. Global firms now frequently have research laboratories located in several countries working on a common project. If regulatory regimes make it excessively difficult to move ideas, equipment, products and people in and out of a United States research facility, the facility can simply be "quarantined" and the work performed abroad. In such cases, prototype shops often follow the research effort . . . and then factories—along with the jobs they all provide. A similar circumstance exists when seeking to sell products outside the United States that have been jointly developed/produced by a United States and foreign firm.

"Deemed exports" can, for example, require a professor at a United States university, if there is a foreign-national student in the class, to obtain an export license before discussing material that is not even classified. The penalties for failure to comply with these relatively arcane laws can be severe.

Several independent reviews have suggested that the export laws be rewritten and focused upon that smaller set of potentially highly sensitive issues (such as nuclear weapon technology, toxins, and the like) rather than seek to apply constraints to items that can be commonly and openly purchased elsewhere in the world or are of lesser consequence (e.g., handcuffs, shotguns, and the like). The 2009 National Research Council report, *Beyond Fortress America* recommends a restructuring of United States export controls to better balance U.S. and national security interests by establishing a one-year sunset provision, subject to renewal, on the listing of protected items. [64]

Visa Policy

Although substantially streamlined since the tightening that immediately followed the events of 9/11, the number of individuals with needed skills who can be admitted to the

---

[64] National Research Council, *Beyond 'Fortress America': National Security Controls on Science and Technology in a Globalized World*, National Academies Press, 2009. Available at: *http://www.nap.edu/catalog.php?record_id=12567*.

United States under the legal quota has been markedly reduced in more recent years. For example, the limit on the number of H-1B visas granted annually was reduced by two-thirds to 65,000 (0.02 percent of the United States population) in 2003 after temporary increases expired.[65] Even in the economic downturn of 2010, H-1B visa quotas were reached months before the end of the application period.[66] These limitations on much-needed talent are in spite of an estimated twelve million illegal immigrants currently residing within the nation's borders.[67] The barriers that scientists and engineers holding temporary visas face in obtaining green cards further affects America's ability to attract and retain a share of the world's "best and brightest" from abroad.

Availability of Markets

In many industries, companies have traditionally tended to locate prototype shops and at least some serial production near R&D facilities. Likely increases in the cost of transportation due to energy price increases—attributable to market conditions, carbon taxes or recovery costs—may motivate firms to conduct manufacturing activities closer to their customer base. China is projected by some to become the largest consumer market in the world in the next decade.[68] By 2025 India's middle class is expected to grow from today's 50 million citizens to just under 600 million.[69]

Employment Policy

The United States has more intrusive employment policies pertaining to such matters as termination rights, minimum wages, unionization, etc., than most developing nations, but considerably less demanding than those of Europe. Many of these policies perform important functions, such as protecting worker health and safety; others, particularly those entailing massive reporting activities, can be counterproductive to a nation's competitiveness.

---

[65] "H-1B visas" are non-immigrant visas that allow U.S. firms to temporarily employ foreign workers possessing special skills.

[66] U.S. Citizenship and Immigration Services, USCIS Reaches FY 2010 H-1B Visa Cap, December 22, 2009.

[67] J.S. Passel and D. Cohn, *A Portrait of Unauthorized Immigrants in the United States*, Pew Hispanic Center, April 14, 2009.

[68] Q. Xiao, China May Be Biggest Consumer Market by 2015, China Daily, February 25, 2010. Available at: *http://www.chinadaily.com.cn/china/2010-02/25/content_9504280.htm*.

[69] D. Farrell and E. Beinhocker, Next Big Spenders: India's Middle Class, *BusinessWeek*, May 28, 2007.

Stability and Predictability of Government, Markets, etc.

The United States enjoys a distinct advantage over most developing nations with respect to government stability and predictability and is at least on a par with most developed nations in this regard. Some recent events notwithstanding, consumer and financial markets, at least when compared with those of some other nations, are remarkably reliable and transparent. In terms of "Overall Risk of Doing Business," the United States has the 13th most favorable rating of 186 nations according to *Euromoney*.[70]

Availability of Transportation and Telecommunications

In most forms of transportation the United States is well served (e.g., 4.0 million miles of roads vs. 2.2 million in China; 15,095 airports vs. 482 in China, 349 in India and 1,216 in Russia).[71] However, as previously noted, this is not the case in broadband telecommunications where, in terms of density, the United States ranks 22th among the world's nations.[72]

Market Growth Potential

As discussed elsewhere, there can be important business advantages—although decreasing as global mobility increases—to be derived from locating factories near potential customers; engineering facilities near factories; and research laboratories near engineering facilities. Given the huge United States consumer market throughout most of the twentieth century, this has been a fundamental competitiveness discriminator possessed by United States firms. However, the United States is now a relatively mature market, the size advantage of which is being eroded or even eclipsed as a large middle-class evolves in developing nations. Speaking of Wal-Mart's shifting focus toward nations other than the United States, Dean Junkans, chief investment officer for PNC Wealth Management, noted that, "The U.S. consumer is tired" . . . not to mention outnumbered.[73] It is estimated that within a decade 80 percent of the world's middle-class will reside in what are now

---

[70] Country Risk 2010: A Fragile Sense of Stability, *Euromoney*, March 2010.

[71] See CIA World Factbook, available at: *https://www.cia.gov/library/publications/the-world-factbook/index.html*.

[72] S. Dutta and I. Mia, *Global Information Technology Report 2009–2010: ICT for Sustainability*, World Economic Forum, 2010.

[73] Y.Q. Mui, As growth in the U.S. slows, Wal-Mart puts more emphasis on foreign stores, *The Washington Post*, June 8, 2010.

categorized as developing nations. There are already 80 million people in China who can reasonably be characterized as middle-class.[74] Globally, it is estimated that by the mid-2020's, there will be two billion such consumers—with the number in China exceeding the total population of the United States at that time by a factor of two.[75] It has been estimated that by 2030, two billion people will join the world's middle class, with most of the addition coming from what are now considered developing countries.[76] By 2020, 70 percent of China's population is expected to have reached middle class status.

Summary

A large number of factors, mostly controlled by government, can strongly impact a nation's ability to create jobs for its citizens in a competitive marketplace. While possessing many inherent advantages because of its democracy and free enterprise system, America also has noteworthy disadvantages—*many of which are self-imposed.*

---

[74] A. Hodgson, China's middle class reaches 80 million, *Euromonitor*, July 25, 2007.

[75] China's middle class population could total 700 million by 2020, *People's Daily*, July 20, 2010.

[76] D. Wilson and R. Dragusanu, *The Expanding Middle: The Exploding World Middle Class and Falling Global Inequality*, Goldman Sachs Global Economics Paper No. 170, July 7, 2008. Note that for Wilson and Dragusanu, those with per capita incomes in the $6,000 to $30,000 range (adjusted for purchasing power) are considered middle class.

5.0
A Category 5 Storm

A TALE OF TWO JOB SEEKERS

It is instructive to ask which of the following two job candidates one would hire:

Candidate "A," ranks in the lower quartile of the high school class, expects to be paid a wage of $17 per hour (the lifetime average wage of a United States high school graduate) with an additional one-third of that amount in benefits.[1] Candidate "B" speaks two languages fluently, ranks near the top of the class and is eager to work for $1.50 per hour.

This scenario, although oversimplified, is nonetheless a reasonable representation of the challenge faced by the average United States high school graduate seeking a job in the global job market—setting aside altogether the one-quarter of United States youths who have not received a high school diploma by the time their class graduates.[2]

Members of the boards of directors of corporations increasingly face a corresponding decision when considering where a new factory, logistics center, maintenance facility or research laboratory is to be located. The members of the *Gathering Storm* committee have collectively participated in thousands of board meetings and have observed that with increasing frequency the decision is made, usually very reluctantly, to go abroad. Initially

---

[1] For average lifetime annual salary, see J.C. Day and E.C. Newberger, *The Big Payoff: Educational Attainment and Synthetic Estimates of Work-Life Earnings*, U.S. Bureau of the Census, July 2002. For average annual hours worked, see Organization for Economic Cooperation and Development, *OECD Employment Outlook*, Paris, 2003.

[2] National Center for Education Statistics, *The Condition of Education 2010* (NCES 2010-028), Figure 18-2.

such choices primarily affected assembly workers, but as time has passed the migration impacted back-office administrative workers, logisticians, engineers and, more recently, researchers, architects, and accountants . . . among a growing list of threatened skills. If the growth markets are abroad, research is performed abroad and manufacturing is abroad, "U.S." companies can still prosper financially, shareholders can benefit, and management can be rewarded...but no jobs are created domestically. While there are recent examples of United States firms repatriating some activities that had previously been moved overseas, particularly those requiring close management controls, the number of such jobs is overwhelmed by the number moving or being created elsewhere by U.S. firms.

The *Gathering Storm* report concluded that, "Market forces are *already at work* moving jobs to countries with less costly, often better-educated and highly motivated workforces, and more friendly tax policies." From a shareholder's perspective, a solution to America's competitiveness shortfall has already been found—but it is at the expense of those seeking employment here at home. This represents a major dislocation of interests and loyalties that has as yet not been widely addressed or in many cases even recognized. The overwhelming body of United States law provides little latitude for management to stray beyond the economic interests of a publicly-owned firm's shareholders when making decisions. Similarly, generally accepted economic principles do not support protectionism. But as Ralph Gomory, former senior vice president for science and technology at IBM, stated, ". . . what is good for America's global corporations is no longer necessarily good for the American people."[3]

TRENDS IN COMPETITIVENESS

From America's perspective, events that have occurred over the past five years have both positively and negatively impacted the nation's competitiveness stature. On the positive side, there is a much greater awareness of the peril implicit in continuing in the direction the nation has been drifting for several decades. This is a non-trivial development, given that the basic nature of the competitiveness challenge does not lend itself to any sudden "wake-up call"—such as was provided by Pearl Harbor, Sputnik or 9/11. Also on the positive side of the ledger are past actions that have been taken by the federal government, particularly as part of the American Recovery and Reinvestment Act of 2009. Perhaps of even greater import, a number of states have undertaken their own

---

[3] R.E. Gomory, Testimony to the Committee on Science and Technology, U.S. House of Representatives, June 12, 2007. Available at: *http://democrats.science.house.gov/Media/File/Commdocs/hearings/2007/full/12jun/gomory_testimony.pdf.*

Gathering Storm assessments with findings that echo those from the study conducted by the National Academies—and in some cases the states have followed their findings with concrete actions.[4]

Unfortunately, a number of adverse developments with regard to the nation's competitiveness have also occurred. Prominent among these has been the economic collapse triggered by the proliferation of sub-prime mortgages. Although not rooted in the same fundamental practices as the economic reversal described in the *Gathering Storm* report, the fallout from this relapse has further weakened America's ability to respond to the long-term challenges it faces—including those addressed in the *Gathering Storm* report.

Further, for the first time in many decades the nation's higher education system is being seriously challenged. This is a consequence of the decline in operating funds attributable to reduced endowments and declining tax revenues. Finally, although no nation has escaped the recent financial crisis unscathed, some have fared better than others and have focused additional sums on competitiveness. For example, last year China sustained an annual real GDP growth rate of 9.1 percent while India and Vietnam achieved 7.4 and 5.3 percent, respectively.[5] The United States real growth rate was a minus 2.6 percent. The abovementioned three foreign countries of course have smaller GDP's than the United States (India, for example, by a factor of four in purchasing power terms). But they also have a lower standard of living to maintain—and new funding sources are being generated, the fruits of which can be relatively quickly allocated as the nation's leadership deems appropriate.

OVERALL ASSESSMENT

In balance, it would appear that overall the United States long-term competitiveness outlook (read jobs) has further deteriorated since the publication of the Gathering Storm *report five years ago.*

Today, for the first time in history, America's younger generation is less well-educated than its parents.[6] For the first time in the nation's history, the health of the younger genera-

---

[4] States that have held *Gathering Storm*-inspired convocations or organized studies include Alabama, Arkansas, California, Maryland, and Michigan.

[5] CIA World Factbook. Available at: *https://www.cia.gov/library/publications/the-world-factbook/index.html.*

[6] National Center for Education Statistics, *National Assessment of Adult Literacy*, 2005. Results showed that the functional literacy of U.S. college graduates declined between 1992 and 2003. Summary available at: *http://nces.ed.gov/NAAL/PDF/2006470_1.PDF.*

tion has the potential to be inferior to that of its parents.[7] And only a minority of American adults believes that the standard of living of their children will be higher than what they themselves have enjoyed.[8] To reverse this foreboding outlook will require a sustained commitment by both individual citizens and by the nation's government…at all levels.

The *Gathering Storm* is looking ominously like a Category 5…and, as the nation has so vividly observed, rebuilding from such an event is far more difficult than preparing in advance to withstand it.

---

[7] B. Kuehn, AHRQ: US Quality of Care Falls Short, *Journal of the American Medical Association* 301(23): 2427-2428, 2009.

[8] J.M Jones, Four in 10 Americans See Their Standard of Living Declining, Gallup, June 9, 2008. Available at: *http://www.gallup.com/poll/107749/Four-Americans-See-Their-Standard-Living-Declining.aspx.*

"Second only to a weapon of mass destruction detonating in an American city, we can think of nothing more dangerous than a failure to manage properly science, technology and education for the common good. . . "

United States Commission on National Security for the 21st Century, 2001

APPENDIX A
Some Perspectives

KNOWLEDGE CAPITAL

- "Science is more essential for our prosperity, our security, our health, our environment, and our quality of life than it has ever been."
  President Barack Obama

- "Where nations once measured their strength by the size of their armies and arsenals, in the world of the future knowledge will matter most."
  President Bill Clinton

- "Basic scientific research is scientific capital."
  Vannevar Bash

- ". . . in today's integrated and digitized global market, where knowledge and innovation tools are so widely distributed. . . . : Whatever can be done, will be done. The only question is will it be done *by* you or *to* you."
  Thomas L. Friedman, Author, "The World Is Flat"

- "Technology has been paying the bills in this country . . . we're killing the goose that laid the golden eggs."
  Stan Williams, Senior Fellow, Hewlett-Packard

- "(Without a change in U.S. government policy) the next big thing will not be invented here. Jobs will not be created here. And wealth will not accrue here."

  Paul Otellini, CEO, Intel Corporation

- "We can already see the signs. Major drug companies such as Merck and Eli Lilly used to outsource much of their manufacturing to India and China: now they also outsource much of their research and engineering."

  Walter Isaacson, Former Managing Editor of TIME

- " . . . we are killing engineering . . . it's our future and we are throwing it down the drain."

  Craig Barrett, Retired Chairman & CEO, Intel Corp.

- "No country can lead in today's world unless it leads in science."

  Speaker Nancy Pelosi

- "Nanotechnology is an activity for which this government will not spare money."

  Vladimir Putin, Prime Minister of Russia

- " . . . the value of Faraday's work today must be higher than the capitalization of all shares on the stock exchange."

  Margaret Thatcher, former Prime Minister, U.K.

- "It's time we once again put science at the top of our agenda and work to restore America's place as the world leader in science and technology."

  President Barack Obama

- "The greatest long-term threat to U.S. national security is not terrorists wielding a nuclear or biological weapon, but the erosion of America's place as a world leader in science and technology."

  Gordon England, Former Deputy Secretary of Defense

- "So many of the great inventions in the past in the United States came from labs like IBM and Bell Labs and so on. They are all disappearing . . . "

  Tony Tan, former Deputy Prime Minister of Singapore

- "America's economy is in crisis. We can either drown under the weight of the problem, or we can surf the wave of opportunity that it brings—to put science, engineering and innovation back in their rightful place in our economy."
  Senator Edward Kaufman, Delaware

- "In terms of technological innovation and industrial creativity, the United States is not today the country it once was."
  N. J. Slabbert, Co-author, "Innovation, The Key to Prosperity"

- "Ideas are the new currency in a worldwide knowledge economy."
  Ben Wildavsky, Senior Fellow, Kauffman Foundaiton

- "The dog food industry spends more on R&D than the electrical (power) sector does."
  Alex Kingsbury, U. S. News and World Report

- "Currently, China, Japan and South Korea are far outpacing the United States in manufacturing and producing the clean energy technologies that will underpin a new wave of economic growth."
  Robert Atkinson et al., "Rising Tigers, Sleeping Giants"

- "If we're number one in technology, why do I have to call India for tech support?"
  Jay Leno, Entertainer

- "The search for talent and knowledge has gone global at a dizzying rate."
  William E. Kirwan, Chancellor, University System of Maryland

- "For decades the United States has enjoyed unquestioned leadership in various technologies required for military superiority. This is no longer true."
  Richard Roca, Director, Johns Hopkins University Applied Physics Laboratory

- "The history of modernization is in essence a history of scientific and technological progress. Scientific discovery and technological inventions have brought about new civilizations, modern industries, and the rise and fall of nations . . . I firmly believe that science is the ultimate revolution."
  Wen Jiabao, Premier, People's Republic of China

- "My partners and I found the best fuel cells, the best energy storage, and the best wind technologies were all born outside the United States…we need to restock the cupboard or be left behind."
  John Doerr, Partner, Kleiner Perkins

HUMAN CAPITAL

- "If a nation expects to be ignorant and free, in a state of civilization, it expects what never was and never will be."
  Thomas Jefferson

- "When I compare our high schools to what I see when I'm traveling abroad, I'm terrified for our workforce of tomorrow."
  Bill Gates, Founder, Microsoft Corp.

- "If you don't solve (the K-12 education problem), nothing else is going to matter all that much."
  Alan Greenspan, former Chairman, Federal Reserve

- " . . . we need young people—a smart kid coming out of school—instead of wanting to be an investment banker, we need them to decide they want to be an engineer, they want to be a scientist, they want to be a doctor or teacher."
  President Barack Obama

- "If companies were run like many (K-12) education systems, they wouldn't last a week."
  Thomas Donohue, President, United States Chamber of Commerce

- "We go where the smart people are. Now our business operations are two-thirds in the United States and one-third overseas. But that ratio will flip over (in) the next ten years."
  Howard High, Spokesperson, Intel Corporation

- "Talent will be the oil of the 21st century."
  Deborah Wince-Smith, President, Council on Competitiveness

- "It's not just that kids need to go to school, they need to learn in school."
  Emiliana Vegas, Senior Education Economist, World Bank

- "The future of our nation and people depends not on just how well we educate our children generally, but on how well we educate them in mathematics and science specifically."
  Senator John Glenn, Ohio

- "I think we are in an education arms race with the rest of the world because knowledge will drive job creation. High wage jobs are only going to be created by people who can acquire knowledge."
  Former Florida Governor Jeb Bush

- "I know that talking about (K-12) standards can make people nervous, but the notion that we have 50 different goal posts is absolutely ridiculous."
  Arne Duncan, United States Secretary of Education

- "The growing education deficit is no less a threat to our nation's long-term well-being than the current fiscal crisis."
  Gaston Caperton, President of the College Board

- "This (K-12 performance) is a prescription for economic decline, because we know that the countries that out-teach us today will outcompete us tomorrow."
  President Barack Obama

- "Every minute we wait, we fall further behind other countries."
  Tom Luce, CEO, National Math and Science Initiative

- "The average pay of major college football coaches now stands at more than one million dollars . . . the average yearly salary for an engineering professor is $107,134 . . . and the average math professor earns $81,818."
  Lyric Winik, Author

- "In a global economy, the best jobs are not going to go to the best in your class but to the best in the world. Some of the Asian countries are just outstanding in math and science achievement, and we're way behind."
  Gary Phillips, Chief Scientist, American Institutes for Research

- " . . . [many states are] lying to children and their parents, because states have dumbed down their standards."
  Arne Duncan, United States Secretary of Education

- " . . . a good education is no longer just a pathway to opportunity—it is a prerequisite."
  President Barack Obama

- " . . . the auto industry and our school system were built for another era . . . neither can compete in the 21st century."
  Michael Bloomberg, Mayor, New York City

- "We expect parents to work in the best interest of the kids. We're working in the best interest of the teachers."
  David Spohn, President, Hudson (Ohio) Education Association

- "If present conditions continue, the highest demographic shift in the 21st century will be in Asia—which will soon have the best engineers and doctors—and more specifically, China."
  The Dilenschneider Group

- "When someone tells you that 'Oh, math is not really my thing,' respond back, 'and working at McDonald's isn't mine.'"
  Danny Crichton, Stanford University Student

- "A decade ago, only one in a hundred leading Chinese scientists in the United States would have considered returning (to China). Today, half would."
  Rao Yi, Dean of Life Sciences, Peking University

- "The way we prepare students has barely changed in 100 years."
  Bill Gates, Founder, Microsoft Corporation

- "Since 1995 the average mathematics score for fourth-graders jumped eleven points. At this rate we catch up with Singapore in a little over eighty years . . . assuming they don't improve."
  Norman R. Augustine, Retired Chairman & CEO, Lockheed Martin Corp.

- "Every year 1.2 million students leave school, condemning themselves to a life of poverty and hardship. There are no good jobs for high school drop-outs. Indeed, there's nothing much for high school grads."

  Arne Duncan, United States Secretary of Education

- "College is more valuable to the future economy than petroleum."

  Gregg Easterbrook, Author

- "An investment in education, designed to improve and increase students' skills, is the best and most effective strategy for stimulating economic recovery."

  Professor Eric Hanushek, Stanford University

- ". . . it's hard to see a reason for defending the status quo—less than five percent of Detroit's fourth- and eighth-graders were deemed proficient in math on the most recent (national) test."

  Michael van Beeck, Author

- "We can no longer accept the bottom third of students becoming our teachers and expect successes."

  Bill Brock, former Secretary of Labor

- "You don't hear many kids talking about being engineers."

  Jim Whaley, President, Siemens Foundation

- "Your top 10 percent—your top 30,000 teachers—would be among the best in the world. Your bottom 30,000 should find another profession. And no one in this room can tell me who is in what category. That is a real problem."

  Arne Duncan, United States Secretary of Education

- " . . . when you need to hire 200 engineers, where do you find them? Well, we know where you find them—you find them in India, Russia, Singapore and China."

  Keith Devlin, Stanford University

- "I find myself hiring talent for my companies abroad, not because I want to but because I can't find qualified engineers and scientists in America."

  Larry Bock, Entrepreneur

- "Nearly two-thirds of low-performing schools in 1989 were still low performers two decades later."
  Tom Loveless, Brookings Institution

- "All of us are going where the high IQ's are."
  Bill Gates, Founder, Microsoft Corporation

- "We had more sports-exercise majors graduate than electrical engineering graduates last year. If you want to become the massage capital of the world, you're well on your way."
  Jeff Immelt, CEO, General Electric Co.

- "Folks, we're in trouble. If we were in private business, we'd be out of business."
  Kenneth Burnley, CEO of the Detroit Public Schools

- "The United States now ranks 25th among 30 industrialized countries in math. If I told you your basketball team finished in 25th place, you'd be outraged."
  Robert Wise , Former Governor, West Virginia

- ". . . contrary to . . . the impression that we exist in a world in which 'taxpayers are willing to continually increase subsidies for higher education,' state support of public universities nationally, on a per-student basis, has been declining for more than two decades and was at the lowest level in 25 years even before the current economic crisis. This comes at a time when demand for higher education continues to increase."
  William E. Kirwan, Chancellor of the University of Maryland System

- ". . . of the seventy finalists in an NSA (National Security Agency)-backed computer coding contest, twenty were from China, ten from Russia and two from the USA."
  Patrick Thibodeau, Computerworld

- "The fate of empires depends on how they educate their children."
  Aristotle

THE INNOVATION ENVIRONMENT

- "... the great Silicon Valley innovation machine hasn't been creating many jobs of late—unless you're counting Asia, where American tech companies have been adding jobs like mad for years."
  Andy Grove, former Intel CEO

- "Will America lead ... and reap the rewards? Or will we surrender that advantage to other countries with clearer vision?"
  Susan Hockfield, President, MIT

- "We operate 144 plants in 80 countries producing the full line of P&G's products and, until very recently, *none* of our top-rated plants has been located in the United States"
  R. Keith Harrison, Global Product Supply Officer, Procter & Gamble

- "Patents are only important if you plan on being successful."
  Sherman McCorkle, President, Technology Venture Corporation

- "United States [helicopter] firms are increasingly finding themselves at a disadvantage in the world market when they have to compete against the all-new products of their non-United States competitors."
  Forecast International

- "At this rate ... we'll be buying most of our wind generators and photovoltaic panels from China."
  Arden Bement, former director, National Science Foundation

- "If the United States doesn't get its act together, DuPont is going to go to the countries that do."
  Chad Holliday, Retired Chairman and CEO, DuPont Corporation

- "We educate the best and the brightest and then we don't give them a green card."
  Michael Bloomberg, Mayor, New York City

- "Providing *short-term* recovery packages to the economy without also fixing the fundamental problems, such as education and investment in research, is like giving a candy bar to a diabetic."

  Norman R. Augustine, Retired Chairman & CEO, Lockheed Martin Corp.

- "[We have] immigration policies that have our colleges educating the world's best scientists and engineers and then, when these foreigners graduate, instead of stapling green cards to their diplomas, we order them to go home and start companies to compete against ours."

  Thomas L. Friedman, Author, "The World Is Flat"

- "*Here*, you see, it takes all the running you can do to keep in the same place. If you want to get somewhere else, you must run at least twice as fast as that."

  The Red Queen, "Through the Looking Glass"

A CATEGORY 5 STORM

- "Where is Sputnik when we need it?"

  Bill Gates, Founder, Microsoft Corporation

- ". . . our present crisis is not just a financial meltdown crying out for a cash injection. We are in much deeper trouble. In fact, we as a country have become General Motors—as a result of our national drift. Look in the mirror: G.M. is us."

  Thomas L. Friedman, Author, "The World Is Flat"

- "Hundreds of thousands of manufacturing jobs have gone away, and they are not coming back."

  Yash Gupta, Dean, Johns Hopkins Carey Business School

- "We're better off taking lots of [employees] and moving them out of the United States as opposed to keeping them inside the United States"

  Steven Ballmer, CEO, Microsoft Corporation

- "Years ago when I lived in California we used to say that California was twenty years ahead of the rest of the nation. I fear we may have been right."
  Norman R. Augustine, Retired Chairman & CEO, Lockheed Martin Corp.

- ". . . the time for a remedy that puts Americans back to work, jump-starts our economy and invests in lasting growth is now."
  President Barack Obama

- "A gradual flow of economic power from West to East has turned into a flood."
  The New York Times

- " . . . good jobs are disappearing faster than bad jobs."
  Damon Silvers, AFL-CIO Policy Director

- "We do not know what people who have been displaced are actually going to do in the next five to ten years."
  Ali Velshi, Chief Business Correspondent, CNN

- "They [other nations] have taken this recession [as an opportunity], not to talk about it, not to debate it, but to actually take steps . . . We must do exactly the same thing."
  Chad Holliday, Retired Chairman and CEO, DuPont Corporation

- "If we spend one trillion dollars on a stimulus and just get better highways and bridges—and not a new Google, Apple, Intel or Microsoft—your kids will thank you for making it so much easier for them to commute to the unemployment office. . . . "
  Thomas L. Friedman, Author, "The World Is Flat "

- "[Outsourcing] . . . is not just a passing fancy . . . if you can find high quality talent at a third of the price, it's not too hard to see why you do this."
  Ron Rittenmeyer, Chairman of the Board, EDS

- "We're well on our way to becoming America, the land of the free and the home of the unemployed."
  Norman R. Augustine, Retired Chairman & CEO, Lockheed Martin Corp.

- "The 19th century belonged to England, the 20th century belonged to the United States, and the 21st century belongs to China. Invest accordingly."
  Warren Buffett

APPENDIX B
REPORT REVIEWERS

We wish to thank the following individuals for their review of this report: John F. Ahearne, Sigma Xi; Bruce M. Alberts, University of California, San Francisco; John H. Gibbons, Resource Strategies; William H. Joyce, Advanced Fusion Systems, LLC; W. Carl Lineberger, University of Colorado; Gilbert S. Omenn, University of Michigan; Kenneth I. Shine, University of Texas System; John Brooks Slaughter, University of Southern California; Morris Tanenbaum, AT&T Corporation (retired); and Sheila E. Widnall, Massachusetts Institute of Technology. Although the reviewers listed above have provided many constructive comments and suggestions, they were not asked to endorse it, and did not see the final draft of the report before its release. The review of this report was overseen by Charles M. Vest, President of the National Academy of Engineering. Responsibility for the final content of the report rests entirely with the authors.

APPENDIX C
PROJECT STAFF

Tom Arrison, Senior Staff Officer, Policy and Global Affairs

Guru Madhavan, Program Officer, Committee on Science, Engineering, and Public Policy, and Committee on Science, Technology and Law, Policy and Global Affairs

Neeraj Gorkhaly, Research Associate, Committee on Science, Engineering, and Public Policy, and Board on Global Science and Technology, Policy and Global Affairs

Laura J. Ahlberg, Consultant Editor

APPENDIX D

BIBLIOGRAPHY

SELECTED PUBLICATIONS RELATED TO *RISING ABOVE THE GATHERING STORM* THAT HAVE APPEARED SINCE 2005

Association of American Colleges and Universities. 2007. *College Learning for the New Global Century: A Report from the National Leadership Council for Liberal Education and America's Promise*. Washington, DC. Available at: http://www.aacu.org/leap/documents/GlobalCentury_final.pdf

Arrison, T., Editor. 2009. *Rising Above the Gathering Storm Two Years Later: Accelerating Progress Toward a Brighter Economic Future. Summary of a Convocation*. Washington, DC: National Academies Press. Available at: http://www.nap.edu/catalog.php?record_id=12537

Atkinson, R. 2010. *Eight Ideas for Improving the America COMPETES Act*. Washington, DC: Information Technology and Innovation Foundation. Available at: http://www.itif.org/files/2010-america-competes.pdf

Augustine, N.R. 2007. *Is America Falling Off the Flat Earth?* Washington, DC: National Academies Press. Available at: http://www.nap.edu/catalog.php?record_id=12021

Bond, J., M. Schaefer, D. Rejeski and R. Nichols. 2008. *OSTP 2.0: Critical Upgrade*. Washington, DC: Woodrow Wilson Center for International Scholars. Available at: http://wilsoncenter.org/news/docs/OSTP%20Paper1.pdf

California Council on Science and Technology. 2009. *Creating a Well-prepared Science, Technology, Engineering and Mathematics (STEM) Workforce: How Do We Get from Here to There?* Sacramento, CA. Available at: http://www.ccst.us/publications/2009/2009CalTAC.pdf

California Council on Science and Technology. 2007. *Shaping the Future: California's Response to "Rising Above the Gathering Storm."* Sacramento, CA. Available at: http://www.ccst.us/publications/2006/GSTFrecs.pdf

College Board. 2008. *Coming to our Senses: Education and the American Future, Report of the Commission on Access, Admissions and Success in Higher Education*. New York, NY. Available at: http://professionals.collegeboard.com/profdownload/coming-to-our-senses-college-board-2008.pdf

Council on Competitiveness. 2007. *Competitiveness Index: Where America Stands.* Washington, DC. Available at: http://www.compete.org/images/uploads/File/PDF%20Files/Competitiveness_Index_Where_America_Stands_March_2007.pdf

Educational Testing Services. 2007. *America's Perfect Storm: Three Forces Changing Our Nation's Future.* Princeton, NJ. Available at: www.ets.org/stormreport

Ezell, S. and R. Atkinson. 2008. RAND's Rose-Colored Glasses: How RAND's Report on United States Competitiveness in Science and Technology Gets it Wrong. Washington, DC: Information Technology and Innovation Foundation. Available at: http://www.itif.org/files/2008-RAND%20Rose-Colored%20Glasses.pdf

Hachigian, N. and M. Sutphen. 2008. *The Next American Century: How the United States Can Thrive as Other Powers Rise.* New York, NY: Simon and Shuster.

Kao, J. 2007. *Innovation Nation: How America Is Losing Its Innovation Edge, Why It Matters, and What We Can Do to Get It Back.* New York, NY: Free Press.

Innovation Council (Canada). 2008. *State of the Nation: Canada's Science, Technology and Innovation System.* Ottawa, Ontario: Available at: http://www.stic-csti.ca/eic/site/stic-csti.nsf/eng/h_00011.html

Kauffman Foundation. 2007. *On the Road to an Entrepreneurial Economy: A Research and Policy Guide.* Kansas City, MO. Available at: http://www.kauffman.org/uploadedFiles/entrepreneurial_roadmap_2.pdf

Kuenzi, J. 2008. *Congressional Research Service Report for Congress, Science, Technology, Engineering, and Mathematics (STEM) Education: Background, Federal Policy, and Legislative Action.* Washington, DC. Available at: http://www.sdsa.org/resources/publications/Congressional%20report.pdf

Lewis, J. 2005. *Waiting for Sputnik: Basic Research and Strategic Competition.* Washington, DC: Center for Strategic and International Studies. Available at: http://csis.org/files/media/csis/pubs/051028_waiting_for_sputnik.pdf

Lowell, B.L., H. Salzman, H. Bernstein, and E. Henderson. 2009. *Steady as She Goes? Three Generations of Students through the Science and Engineering Pipeline.* Paper presented at Annual Meetings of the Association for Public Policy Analysis and Management, Washington, DC. Available at: http://policy.rutgers.edu/faculty/salzman/SteadyAsSheGoes.pdf

McKinsey & Company. 2009. *The Economic Impact of the Achievement Gap in America's Schools.* New York, NY. Available at: http://www.mckinsey.com/App_Media/Images/Page_Images/Offices/SocialSector/PDF/achievement_gap_report.pdf

Matthews, C. 2008. *Congressional Research Service Report for Congress, Science, Engineering, and Mathematics Education: Status and Issues.* Washington, DC. Available at: https://www.policyarchive.org/bitstream/handle/10207/19771/98-871_20080328.pdf?sequence=3

Millennium Project. 2008. *Engineering for a Changing World: A Roadmap to the Future of Engineering Practice, Research, and Education.* Ann Arbor, MI: University of Michigan. Available at: http://milproj.dc.umich.edu/publications/EngFlex_report/download/EngFlex%20Report.pdf

National Academy of Engineering. 2008. *The Offshoring of Engineering: Facts, Unknowns, and Potential Implications,* Washington, DC: National Academies Press. Available at: http://www.nap.edu/catalog.php?record_id=12067

National Academy of Engineering and National Research Council. 2009. *Engineering in K-12 Education: Understanding the Status and Improving the Prospects.* Washington, DC: National Academies Press. Available at: http://www.nap.edu/catalog.php?record_id=12635

National Academy of Sciences, National Academy of Engineering, and Institute of Medicine. 2008. *Science and Technology for America's Progress: Ensuring the Best Presidential Appointments in the New Administration,* Washington, DC: National Academies Press. Available at: http://www.nap.edu/catalog.php?record_id=12481

National Research Council. 2007. *Enhancing Productivity Growth in the Information Age: Measuring and Sustaining the New Economy*. Washington, DC: National Academies Press. Available at: http://www.nap.edu/catalog.php?record_id=11823

National Research Council. 2007. *The Future of United States Chemistry Research: Benchmarks and Challenges*. Washington, DC: National Academies Press. Available at: http://www.nap.edu/catalog.php?record_id=11866

National Research Council. 2007. *India's Changing Innovation System: Achievements, Challenges, and Opportunities for Cooperation*. Washington, DC: National Academies Press. Available at: http://www.nap.edu/catalog.php?record_id=11924

National Research Council. 2007. *Science and Security in a Post 9/11 World: A Report Based on Regional Discussions Between the Science and Security Communities*. Washington, DC: National Academies Press. Available at: http://www.nap.edu/catalog.php?record_id=12013

National Research Council. 2008. *Science Professionals: Master's Education for a Competitive World*. Washington, DC: National Academies Press. Available at: http://www.nap.edu/catalog.php?record_id=12064

National Research Council. 2009. *21st Century Innovation Systems for Japan and the United States: Lessons from a Decade of Change*. Washington, DC: National Academies Press. Available at: http://www.nap.edu/catalog.php?record_id=12194

National Research Council. 2009. *Beyond 'Fortress America': National Security Controls on Science and Technology in a Globalized World*. Washington, DC: National Academies Press. Available at: http://www.nap.edu/catalog.php?record_id=12567

National Mathematics Advisory Panel. 2007. *The National Mathematics Advisory Panel Preliminary Report*. Washington, DC: U.S. Department of Education. Available at: http://www2.ed.gov/about/bdscomm/list/mathpanel/pre-report.pdf

National Science Board. 2007. *A National Action Plan for Addressing the Critical Needs of the United States Science, Technology, Engineering, and Mathematics Education System*. Arlington, VA: National Science Foundation. Available at: http://www.nsf.gov/nsb/documents/2007/stem_action.pdf

National Science Board. 2010. Science and Engineering Indicators 2010. Arlington, VA: National Science Foundation. Available at: http://www.nsf.gov/statistics/seind10/

Olson, S. and J.B. Labov, Editors. 2008. *State Science and Technology Policy Advice: Issues, Opportunities, and Challenges*, Washington, DC: National Academies Press.
Available at: http://www.nap.edu/catalog.php?record_id=12160

Panel to Review the National Innovation System (Australia). 2008. *Venturous Australia*. North Melbourne, Victoria: Cutler & Company Ply Ltd. Available at: http://www.innovation.gov.au/innovationreview/Documents/NIS_review_Web3.pdf

Royal Society. 2010. *The Scientific Century: Securing Our Future Prosperity*. London. Available at: http://royalsociety.org/The-scientific-century/